Skeptoid 4
Astronauts, Aliens, and Ape-Men

By Brian Dunning

Foreword by John Rennie
Illustrations by Abby Goldsmith

Skeptoid 4: Astronauts, Aliens, and Ape-Men
Copyright 2011 by Brian Dunning
All Rights Reserved.

Skeptoid Podcast ©2006-2012 by Brian Dunning
http://skeptoid.com

Published by Skeptoid Media
Laguna Niguel, CA

First Edition
ISBN: 978-1475205657
Printed in the United States of America

A man should look for what is, and not for what he thinks should be.
Albert Einstein

ACKNOWLEDGEMENTS

The following crack team of researchers provided invaluable assistance with the referencing and further reading suggestions provided at the end of each chapter:

Mike Bast, Grant Black, Mike Bohler, Josh DeWald, Helen Genevere, Derek Graham, Andrea Hope, James Lippard, Kelly Manning, Mark Metz, Cuauhtemoc Moreno, Lee Oeth, Kathy Orlinsky, Tom Schinckel, Thomas Shulich, Adam Waller, Mike Weaver, and Sarah Youkhana.

To Mom – for pushing a small lad in a swing, so that he saw the sky and the world and the universe, and learned to be amazed.

Contents

Foreword: The Deceptions of Uncertainty 1
Introduction 4
1. Network Marketing 7
2. Wi-Fi, Smart Meters, and Other Radio Bogeymen 13
3. The Scole Experiment 20
4. Vaccine Ingredients 26
5. The Baigong Pipes 34
6. The Naga Fireballs 40
7. The Antikythera Mechanism 46
8. Is Barefoot Better? 53
9. Emergency Handbook: What to Do When a Friend Loves Woo 60
10. Martial Arts Magic 67
11. The Bell Island Boom 73
12. Did Jewish Slaves Build the Pyramids? 80
13. Ball Lightning 87
14. The Faces of Bélmez 93
15. The Denver Airport Conspiracy 99
16. Zeitgeist: The Movie, Myths, and Motivations 107
17. The Georgia Guidestones 113
18. Cargo Cults 120
19. The Virgin of Guadalupe 126
20. The Mystery of Pumapunku 134
21. Therapeutic Touch 140

22. Mengele's Boys from Brazil 147
23. Morgellons Disease 154
24. Dinosaurs Among Us 161
25. The Westall '66 UFO 168
26. The Lost Ship of the Desert 176
27. The North American Union 182
28. Attack on Pearl Harbor 188
29. Mozart and Salieri 195
30. The Things We Eat... 202
31. Some New Logical Fallacies 209
32. The Astronauts and the Aliens 217
33. Stalin's Human-Ape Hybrids 224
34. Yonaguni Monument: The Japanese Atlantis 231
35. The Myers-Briggs Personality Test 238
36. Toil and Trouble: The Curse of Macbeth 245
37. The Frog in the Stone 251
38. The Brown Mountain Lights 258
39. Boost Your Immune System (or Not) 266
40. Speed Reading 273
41. DDT: Secret Life of a Pesticide 279
42. The Mystery of STENDEC 286
43. The South Atlantic Anomaly 293
44. IQ Testing 299
45. Whales and Sonar 306
46. Hollywood Myths 313
47. Gluten Free Diets 325

48. Mystery Spots 331
49. The Alien Buried in Texas..337
50. Nuclear War and Nuclear Winter..................................344

Foreword: The Deceptions of Uncertainty

by John Rennie

Here's my tribute to Brian Dunning's intellectual integrity: Within hours of my first real meeting with him, I was publicly correcting him — and he laughed about it.

The scene was New York in the spring of 2011, at a discussion I was leading at the annual Northeast Conference on Science and Skepticism on "Information Overload: How Do We Know Whom to Trust?" Brian was part of the international panel of skeptics gathered for the occasion, which also included Kendrick Frazier, George Hrab, and Sadie Crabtree. People's schedules being busy as they were, I hadn't had a chance to do much more than say hello and talk logistics with any of them before we all found ourselves onstage.

I introduced the talk by citing the wisdom of a saying commonly attributed to Mark Twain: "It ain't what you don't know that gets you into trouble. It's what you know for sure that just ain't so." Any of us can easily be led astray by misinformation we've taken to heart and assumptions we've neglected to question.

The only problem with that quote, I continued, is that there's no evidence Twain ever said or wrote those words. Nor can they be pinned definitively to Will Rogers or any of the other famous wags who sometimes get credited. The earliest citation anyone has found for that sentiment seems to be from a 19th-century newspaper column written by a newspaper editor who may have known Samuel Clemens.

When he heard that, Brian guffawed that the quote, with its Twain attribution, appeared in his then-new book, *Pirates,*

Pyramids, and Papyrus. And then he went on to speak eloquently about the trustworthiness of well-established science.

Brian's response comes as no surprise to anyone who knows him from his Skeptoid podcasts and writings. He stands four-square not just for being right (that is, holding factually accurate views) but for a logical, methodologically sound approach to drawing conclusions. He knows that simply being wrong isn't necessarily a badge of shame—but refusing to be right is.

This collection of Brian's essays is full of such correctives that I wish more of the world would take to heart. As I write this foreword, for example, the early stages of the U.S. presidential race are roiling with the anti-vaccination views of some of the leading contenders. I suspect that some of those politicians are too far gone to benefit from reading Brian's debunking of those worries (see Chapter 4, "Vaccine Ingredients"), but is it too much to hope that some of their constituents might not be?

Perhaps some of them would also then see that when confronted with the mysterious, they needn't rush to embrace paranormal, ahistorical fantasies. Sure, ancient aliens or lost civilizations might at first seem like the only possible explanations for 150,000-year-old metal pipes found in a Tibetan cave... but the more prosaic explanation that has emerged (see Chapter 5, "The Baigong Pipes") is in many ways just as wonderful.

Those readers would also learn that skeptics as smart as Brian Dunning aren't afraid to concede that sometimes science doesn't know all the answers — yet. Case in point: the recent debate about the supposed advantages of running without shoes (see Chapter 8, "Is Barefoot Better?"). Humility can be refreshing! (Or so I'm told; I wouldn't know.)

Making fun of those with irrational beliefs is easy. But that attitude misses what is also so poignant about them. Many of them are smart, perceptive, well-educated folks with backgrounds in science or engineering; their backgrounds just have some unfortunate holes they don't recognize. It's the tragedy of the woo: people driven to madness by their misplaced certainty that they know best. It ain't what they don't know....

Brian Dunning, on the other hand, does know. If he's sure something is true, he'll cheerfully show you why. And in the extremely unlikely event that you can prove that he's wrong, he'll enjoy the moment as much as you.

John Rennie is a science writer, the former editor in chief of Scientific American, *and an advisor to the New York City Skeptics.*

Introduction

It would be inaccurate to call this book a labor of love. It's more like a joyride.

There are fifty chapters here, exploring fifty stories; many of which you've heard, many that you haven't. I knew something about most of them before sitting down to do the research, or at least I thought I did. The lesson for me was that even the simplest of urban legends can hide a wealth of fascinating history, science, and human nature. When you attack any subject with an honest thirst for insight, you nearly always get more than you ever imagined.

Like so many of my fellow science journalists, I spent my youth obsessed with books about science, adventure and science fiction novels, and books on the paranormal: Bigfoot, UFOs, ghosts, the usual. They were so well thumbed I practically had some of them memorized. But there was one thing that I did rather poorly, and that was to analyze what I read with a critical mind.

Anyone who wrote a book, I reasoned, certainly knew what he was talking about; and if it was in print it must be true.

And so I grew up with the firm knowledge that the largest Bigfoot on record was fourteen feet tall (spotted by a troop of Boy Scouts), that Betty and Barney Hill had been abducted and subjected to medical tests by gray aliens, and that one could live forever if he only tuned into the evening news and learned what new toxin caused cancer and could be avoided. Nobody ever suggested to me that there was a different way to look at these stories, and thus I never did so.

But age eventually brought me focus and I discovered the reason I was interested in such things was not because the subject matter was especially appealing, but because to learn about them was to *learn*. And I found that popular tales, glibly reported, were but a cheap gilding – a fast food paper wrapping, made for on-the-go consumption by the masses. Pull that wrapping aside, discard the mass market sound bite, and what lies inside is virgin territory. It's a real story, supported by things we can actually learn through science, that almost nobody gets to see.

People are too busy these days. People don't have time to learn. They want Oprah and the Action News to tell them something they can understand in fifteen seconds. The media delivers this, in spades, and the excitement of actually learning is becoming an endangered species.

If you've listened to my weekly Internet show Skeptoid (at Skeptoid.com), you've probably heard in my voice how genuinely excited I am to have learned the real facts behind some of these stories. We use the power of skepticism and critical thinking to tear away that gilt wrapper, and we lay it bare with the laser beam of the scientific method. Not once have I ever researched a subject where I did not learn something new that thrilled me, and not once in these fifty chapters will you read a story you knew everything about.

We don't know everything. Every one of these chapters is incomplete. Maybe one or more of them will be a starting point for you to enjoy your own expedition into science as I have.

I hope so.

– Brian Dunning

BRIAN DUNNING

1. Network Marketing

Call them Network Marketing, Multilevel Marketing, or MLM, these pyramid plans are proven not to work.

In this chapter, we're going to point our skeptical eye at network marketing plans, formerly known as multilevel marketing or MLM (name changed to escape the stigma). They say that when there's a gold rush, the way to make money is to sell shovels. Network marketing companies sell shovels, along with dreams of gold: All you have to do is go out there and dig, dig, dig, and buy more shovels, and get your friends to buy shovels too. Levi Strauss and other suppliers became millionaires, and hundreds of thousands of miners went broke.

Network marketing plans are started by a company selling some product — fruit juice, soap, vitamin pills, water filters; anything, it doesn't matter — through a network of independent distributors who are promised exponential commissions by recruiting multiple levels of other distributors beneath them. The company is guaranteed sales because the distributors are required to make minimum purchases, on which commissions trickle upward. There's little need to actually go out and try to sell the product to anyone; money is made by building your network of distributors beneath you, and their distributors beneath them. Soon the commissions trickling up from all those monthly purchases combine into a raging torrent of cash. And if you just buy a few more shovels, you're sure to strike gold.

Network marketing plans differ from illegal pyramid schemes only by one subtle point: Commissions can only legally be paid on sales of a physical product. If commissions are offered upon recruitment of new distributors, then it's defined as an illegal pyramid scheme. Pyramids are illegal because they necessarily collapse when nobody else can be recruited. However the illegal plans are pretty rare; most companies are smart enough to stay on the right side of the law. But the problem of community saturation, and inevitable collapse, remains.

A tipoff that should clue you into the wisdom of network marketing is that the companies themselves, who manufacture and sell the product, don't even eat their own dog food. They are making money the old fashioned way, by selling an expensive product. It's *you* whom they recruit to start a network marketing business. When an existing distributor pitches you and gets you to become a distributor yourself, you are required to make your initial purchase of "inventory" of whatever the product is. You either consume that product yourself or sell it to others. Your principal sales tool is the pitch that if your customers become distributors beneath you, they can buy the product at a discounted wholesale price. In most plans, in order to retain your distributor status and qualify for the wholesale discount, regular monthly purchases have to be made.

But even this discounted wholesale price is usually far higher than the market value of comparable products available from the supermarket. Participants nearly always find themselves in the unenviable position of having invested a lot of money in their own required inventory purchases, and desperately trying to recruit new distributors in an effort to earn commissions on *their* inventory purchases, and hopefully recover their own investment. So this raises the question: How often does it work out that way? How many MLM participants ever recover their own investments?

The Federal Trade Commission cautions "Most [network marketers] end up with nothing to show for their money except the expensive products or marketing materials they're pressured to buy."

Consumer Reports advises "Stay away from multilevel marketing schemes that make earnings contingent on your ability to sign up an ever-growing pyramid of 'distributors' who are supposed to do the same and pass sales commissions up the line."

The nonprofit Consumer Awareness Institute analyzed available data published by the MLM companies themselves. Of the companies surveyed, they reported the least successful was Amway/Quixtar where 99.99% of distributors lose money, and the most successful was Herbalife, where 99.42% of distributors lose money.

They also surveyed 200 tax preparers in three counties in Idaho and Utah, where 6% of residents are active network marketing participants. From over 300,000 tax returns, not a single one reported significant profits from network marketing activities.

In a Wisconsin lawsuit, the tax returns of the top 200 of 20,000 network marketing participants were examined by the Attorney General. The average income of this *top 1%* was a loss of $900.

Newsweek found that fewer than 1% of Mona Vie distributors ever qualified for any commission at all, and less than 1 in 1,000 recovered the cost of their required monthly purchases.

So if network marketing plans don't work, why do people buy into them? Network marketing plans are easily sold by simply laying out some compelling mathematics on a whiteboard. A typical program sets five downline members as the goal for each participant: To be successful, you need only recruit enough people to end up with just five who actively participate. Below those five are their five apiece, totaling 25. This is your network. Each downline of five are qualified by participating at the minimum required level, so this model already excludes everyone who is flakey or only half-hearted, leaving only the five good ones in each downline. Your commissions based on those minimum participation levels — where all five below you dutifully make their minimum monthly inventory purchases — guarantees you an impressive income. The mathematics are black and white, and it's so simple that nothing can go wrong. You'd have to be stupid not to do it.

But here is the problem that these whiteboard presentations always manage to omit. *Of all the thousands of network marketing plans available now or in the past, if only one of them had ever had even a single line active to only 14 levels deep, that alone would have required the participation of more human beings than exist.* That math is black and white, too. Level 14 is populated by 5^{14}, or about 6.1 billion people, the entire population of the planet, in addition to level 13 with 1.2 billion, all the way up to you and your original five. You can answer "Oh sure, but a lot of the people don't get all five or they flake somehow," but you forget that the entire premise has already eliminated those who flake or who don't get all five. The unfortunate conclusion is that a fully invested network, upon which the whiteboard presentations are dependent, *has never actually happened.*

A fundamental reason that such networks fail is that they depend upon recruiting people to compete with you. If you own a shoe store, and you pitch every customer on opening their own shoe store instead of being your customer, very soon you're going to have a neighborhood full of shoe stores, with everybody trying to sell and nobody left to buy. It doesn't take an MBA to see that this is pretty much the polar opposite of a sound business strategy.

Let's say you tried to make it sound, and said "Forget the multilevel recruiting, I'm going to focus on selling the product." Is anyone doing that successfully? It would not appear so. During yet another lawsuit in the UK, the government found that less than one in ten participants ever sold even a single product to another person. Since the company has its distributors as a captive audience required to make regular purchases, the products are typically grossly overpriced compared to similar products available in supermarkets. This makes their sale a dubious prospect for those few distributors who ever do attempt retail sales to customers. Surveys show that nearly all products purchased by network marketers are consumed by the distributors themselves.

This fact is rarely mentioned in the sales pitches. Instead, they typically promote the merchandise (referred to as "lotions & potions" by MLM critics) as wondrous super products that will be in high demand. But, you should always beware of success stories coming from MLM distributors. Most MLM companies pay shills who lie about having had multimillion dollar success with the scheme. These are typically the ones who travel around giving seminars, pitching motivational materials, and putting on recruiting extravaganzas that have been criticized by the Federal Trade Commission for promoting an almost cult-like religious mania as a substitute for sound business practices.

I've spoken with enough friends and other people who are into network marketing to know that the default response to this is "Oh, but this plan is different." Sure, every plan has different tweaks and details, but fundamentally they are all the same. The company is going to make tons of money selling an outrageously overpriced product every month to their captive audience buyers: You, and any friends you recruit. Not one of you has any realistic hope of coming out ahead. My advice to everyone involved in network marketing: Simply stop now. Stop convincing yourself that profits are just around the corner if you just buy a few more cases of expensive product. Just stop now, walk away, consider it a lesson well learned, and don't give them another dollar.

One final tidbit I'll leave you with. On average, 99.95% of network marketers lose money. However, only 97.14% of Las Vegas gamblers lose money by placing everything on a single number at roulette. So if you're thinking about joining a network marketing plan, and aren't dissuaded by the facts I've presented, consider instead going to Vegas and placing all your money in a single pile on number 13. Sooner or later you're going to have to take my advice and just stop now.

References & Further Reading

Bloch, Brian. "Multilevel marketing: what's the catch?" *Journal of Consumer Marketing.* 1 Oct. 1996, Volume 13, Issue 4: 18-26.

Coward, C. "How to Spot a Pyramid Scheme." *Black Enterprise.* 1 Feb. 1998, Volume 28, Number 7: 200.

Dokoupil, T. "A Drink's Purple Reign." *Newsweek.* Newsweek Inc., 11 Aug. 2008. Web. 22 Aug. 2008. <http://www.newsweek.com/id/150499/page/1>

FTC. "The Bottom Line About Multilevel Marketing Plans and Pyramid Schemes." *Protecting America's Consumers.* Federal Trade Commission, 1 Oct. 2009. Web. 13 Oct. 2009. <http://www.ftc.gov/bcp/edu/pubs/consumer/invest/inv08.shtm>

Vander, N., Peter, J., Keep, W. "Marketing Fraud: An Approach for Differentiating Multilevel Marketing from Pyramid Schemes." *Journal of Public Policy & Marketing.* 1 May 2002, Volume 21, Number 1: 139-151.

Walsh, J. *You can't cheat an honest man: How Ponzi schemes and pyramid frauds work and why they're more common than ever.* Aberdeen, WA: Silver Lake Publishing, 1998. 183-202.

2. WI-FI, SMART METERS, AND OTHER RADIO BOGEYMEN

Are common radio transmitters carcinogenic or otherwise harmful?

In this chapter, we're going to point the skeptical eye at popular claims that ordinary radios — such as walkie talkies, police and emergency radios, and those embedded in devices such as cell phones, Wi-Fi hubs, and smart utility meters — are dangerous. Some say they cause cancer, some say they present other more nebulous health risks. How concerned do you need to be that something as ubiquitous as radio could be doing you more harm than good?

This issue rose to the headlines in popular media with a frightening announcement in May of 2011 by the World Health Organization. The press release stated that the International Agency for Research on Cancer (IARC) had placed radiofrequency (RF) in their Group 2B of possible carcinogens due to an increased risk of the brain cancer glioma associated with the use of mobile phones. Unfortunately, very few people actually read the release, and saw only that headline, which presents a highly skewed perspective of what was actually said. As a result, new movements arose worldwide, notably in Canada, for certain RF devices to be banned. Canada's Green party openly called for the elimination of Wi-Fi computer networks in schools, and many groups have campaigned against the purported health effects of smart meters.

My question to the groups actively campaigning against stuff that's in Group 2B is "Do you drink coffee?" Most do, and yet coffee is also in Group 2B. So are the crafts of carpentry and joinery. Pickled vegetables, coconut oil, and even the Earth's magnetic field are in Group 2B. Now, granted, it would be fallacious logic to say that just because these other things sound ordinary and safe, that makes radiofrequency safe; but it is true that the World Health Organization considers them to be similarly risky.

- Group 1 is the classification for things that have been found to be carcinogenic. This includes ultraviolet radiation, tobacco, and plutonium.

- Group 2A is the classification for probable carcinogens, things that have not yet been found to cause cancer but for which there is good evidence they might. This includes engine exhaust and working in the petroleum industry.

- Group 2B is the list of possible carcinogens, which are things that have not been found to cause cancer but for which there is cause to study further. It is a list of items which have not — repeat, *not* — been found to be carcinogenic. Will they tomorrow? Maybe, but they're not now, according to what we know so far.

If the World Health Organization is the authority whose word you're going on, then you should look at what they actually say. Their position paper on radio frequencies and electromagnetic radiation states unequivocally that:

> ...*Current evidence does not confirm the existence of any health consequences from exposure to low level electromagnetic fields.*

Nor should we expect such consequences. Radiofrequency is all around us, and always has been. Tune any radio to a station containing static and what you're hearing is normal background radiation. About 1% of that static is actually left over from the Big Bang. But just because radiofrequency is natural for all living beings throughout the universe, that doesn't mean it's safe. To determine whether something is safe, we look at the data. So let's look at what we know so far.

The electromagnetic spectrum is pretty simple to understand. It starts at the low end with a frequency of zero, up through the radio frequencies, past visible light and up through gamma rays and onto infinity, with higher and higher frequencies. The frequencies at the lower end are what we call non-ionizing, because they lack sufficient energy to strip electrons and change chemistry. The frequencies at the higher end are ionizing, which makes it damaging to living tissue. The dividing line between the two is the upper end of visible light, where ultraviolet begins. A sunburn is actually tissue damage caused by ionizing radiation; that UV has enough energy to just barely penetrate the outer layer of your skin. But as we go even higher, into the X-ray range, the radiation is energetic enough to penetrate all the way through your body. X-rays can be stopped by the lead-lined blanket they give you. But even higher energy frequencies, like the strongest cosmic rays, can go all the way through the entire planet.

Smart meter

So remember that dividing line. Visible light, like that inside your home, is generally safe as are all the radio frequencies below it. Ultraviolet light, and everything higher, is damaging.

Yet claims persist of harm from non-ionizing radiation, and they'll often cite studies showing a biological effect from some manifestation of radio. There are only a handful of such studies which are repeatedly cited, in comparison to the more than 25,000 studies surveyed by the WHO that found no reason for concern.

Perhaps the most vocal of all the anti-radio activists is Dr. Magda Havas at Trent University in Ontario, Canada. You'll be hard pressed to find a mass media article about the safety of radio devices that doesn't cite Dr. Havas as its expert. She hasn't published any good research of her own, rather she tirelessly cites these few fringe studies over and over again to promote the idea that radio is harmful. To find such studies, you have to dig past hundreds of studies that contradict her desired results. It's hard to imagine that Dr. Havas is unaware that she's promoting science that's in direct conflict with what virtually everyone else has found. You have to wonder whether her students accept her claims at face value, or whether they view it within the context of the scientific consensus.

Dr. Havas cites one such study that she says showed mobile phone signals break down the blood brain barrier. In fact, this study was a single in-vitro (petri dish) experiment, and the authors only hypothesized that one potential effect might be to increase the blood brain barrier permeability. In other words, nobody has ever observed such an effect.

Another study is often cited as showing that non-ionizing microwaves have been found to cause single and double strand DNA breakage. While this study was interesting, it was very small — only four groups of rats —and has not been replicated by any other researchers. In addition, it exposed the rats to a type of signal not found in either nature or in electronic devices (a powerful, continuous 2.5 GHz tone) and the effects disappeared when the signal was augmented with background noise. The lead author, Dr. Henry Lai, is the co-editor of *Electromagnetic Biology and Medicine*, a journal dedicated to the promotion of alleged biological harm from radio.

The third study the anti-radio activists promote most often is said to show that radio signals increase blood sugar, leading to diabetes. If you're wondering why so many of us live in a radio-soaked world but don't have diabetes yet, the answer lies in the quality of this study. It was Magda Havas' own research, in which she published the self-reported results of four people who identified as being both diabetic and "electrosensitive", and who said they felt better after moving away from their electronic devices. The study has essentially zero scientific validity. What was the only journal that published her article? Dr. Lai's *Electromagnetic Biology and Medicine*.

The sad thing is these researchers missed the boat, because non-ionizing radiation does have at least one real effect on living tissue: heat. This is why you feel warm in sunlight; the sun's gargantuan output completely blows away all the other sources we're exposed to, either manmade or natural. When electromagnetic radiation strikes an absorbent surface, like your skin, that energy is converted into heat. Simple thermodynamics.

If you aim a laser pointer at your hand, you won't feel anything. Some of the energy of that light reflects from the visible spot that we see, and some of it is absorbed by your hand and converted into heat, but there's not nearly enough heat for your nervous system to detect. But crank up the power to that of an industrial laser, and it could burn a hole right through your skin. Turn it back down to the power of a medical laser and it can excise a mole or make a precision cut. This is tissue damage from non-ionizing radiation. The mechanism is simply heat.

There are other medical applications for RF-generated heat at frequencies below that of visible light. RF is also used in dermatology and arthroscopic surgery. The basic idea here is to cook and shrink collagen fibers. This can be used to tighten skin to reduce wrinkles, to shrink ligaments and pull loose joints tight, and to ablate surfaces for cleaning up joints or attacking tumors. Radiofrequency probes are used in arthroscopic surgery, and although they're quite different from a microwave oven, they work on the same principle; but with a far lower wattage (up to about 30 W), and much lower frequency (about 6 MHz). Since these radio frequencies do not penetrate the body to any degree, the probe is placed in direct contact with the ligament. The rapidly oscillating RF field twists the water molecules back and forth, and the surface is heated by the friction, called the dielectric heating effect. This temperature gets as high as about 150°F/65°C. In dermatology the wand is applied to the wrinkled skin and performs similar heating, which would be quite painful and so the wand simultaneously applies a coolant. In a monopolar surgical probe, the heating effect is extremely localized and is limited to the surface in contact with the probe; and in a bipolar probe, the field oscillates between two closely spaced electrodes, and the heating is limited to that small space.

Discounting the heat from the battery or power supply circuitry, why don't we feel any radiated heat from a Wi-Fi hub or a smart meter, or any other familiar radio transmitter? It's because there's not nearly enough power and it's not highly focused like a laser. Television's *Mythbusters* once tested this myth by strapping an uncooked turkey to a ship's high-powered radar antenna and found no measurable heating, just as we'd expect.

One day the science might change and we might learn that there actually is credible cause for concern about radio frequencies. Until it does, enjoy the services that radio provides; and don't forget to try that thing with listening to a static channel on the radio. It's really cool when you understand what you're listening to.

References & Further Reading

ASDS. "Technology report: Monopolar radiofrequency." *American Society for Dermatologic Surgery.* American Society for Dermatologic Surgery, 1 Jan. 2010. Web. 12 Aug. 2011. <http://www.asds.net/TechnologyReportMonopolarRadiofrequency.aspx>

Hecht, P., Hayashi, K., Cooley, A., Lu, Y., Fanton, G., Thabit, G., Markel, M. "The Thermal Effect of Monopolar Radiofrequency Energy on the Properties of Joint Capsule." *American Journal of Sports Medicine.* 1 Nov. 1998, Volume 26, Number 6: 808-814.

IARC. *Agents Classified by the IARC Monographs.* Lyon: World Health Organization, 2011.

IARC. "IARC Classifies Radiofrequency Electromagnetic Fields as Possibly Carcinogenic to Humans." *International Agency for Research on Cancer.* World Health Organization, 31 May 2011. Web. 15 Aug. 2011. <http://www.iarc.fr/en/media-centre/pr/2011/pdfs/pr208_E.pdf>

Novella, S. "CFLs, Dirty Electricity and Bad Science." *Science-Based Medicine.* New England Skeptical Society, 22 Sep. 2010. Web. 15 Aug. 2011. <http://www.sciencebasedmedicine.org/index.php/cfls-dirty-electricity-and-bad-science/>

Shermer, M. "Can You Hear Me Now? The Truth about Cell Phones and Cancer: Physics shows that cell phones cannot cause cancer." *Scientific American.* 4 Oct. 2010, Volume 303, Number 4: 98.

3. The Scole Experiment

Said to be the best evidence yet for the afterlife -- but how good is that evidence?

Turn out the lights and link your hands, for in this chapter we're going to hold a séance and contact the dead, and have them perform parlor tricks for us in the dark. We're going to look at the Scole Experiment, a large, well-organized series of séances conducted by members of the Society for Psychical Research in the late 1990's in Scole, a small village in England. Reported phenomena included ghostly lights flitting about the room, images appearing on film inside secure containers, reports of touches from unseen hands, levitation of the table, and disembodied voices. Due to the large number of investigators and sitters involved, the number and consistency of paranormal episodes observed during the séances, and the lack of any finding of fraud, many believers often point to the Scole Experiment as the best scientific evidence that spirits do survive in the afterlife, and can and do come back and interact with the living, demonstrating an impressive array of conjuring powers.

There were a total of six mediums and fifteen investigators from the SPR. The Society for Psychical Research, or SPR, is based in London and is more than a century old. Its membership consists of enthusiasts of the paranormal. The authoritative source for what happened in the Scole Experiment is a report several hundred pages long, called The Scole Report, originally published in the journal *Proceedings of the Society for Psychical Research*, and written by three of the lead investigators who were present at the sittings, all current or former senior officers of the SPR: plant scientist Montague Keen, electrical engineer Arthur Ellison, and psychologist David Fontana. I have a copy here on my desk. It goes through the history of how the experiments came together, details each of the many séances, and presents analysis and criticism from a number of the SPR investigators who observed.

Unfortunately, the Scole Experiment was tainted by profound investigative failings. In short, the investigators imposed little or no controls or restrictions upon the mediums, and at the same time, agreed to all of the restrictions imposed by the mediums. The mediums were in control of the séances, not the investigators. What the Scole Report authors describe as a scientific investigation of the phenomena, was in fact (by any reasonable interpretation of the scientific method) hampered by a set of rules which explicitly *prevented* any scientific investigation of the phenomena.

The primary control offered by the mediums was their use of luminous wristbands, to show the sitters that their hands were not moving about during the séances. I consulted with Mark Edward, a friend in Los Angeles who gives mentalism and séance performances professionally. He knows all the tricks, and luminous wristbands are, apparently, one of the tricks. There are any number of ways that a medium can get into and out of luminous wristbands during a séance. The wristbands used at Scole were made and provided by the mediums themselves, and were never subjected to testing, which is a gross dereliction of control by the investigators. Without having been at the Scole Experiment in person, Mark couldn't speculate on what those mediums may have done or how they may have done it. Suffice it to say that professional séance performers are not in the least bit impressed by this so-called control. Tricks like this have been part of the game for more than a century. Since hand holding was not employed in the Scole séances, the mediums effectively had every opportunity to be completely hands free and do whatever they wanted to do.

Believers in the Scole Experiment are likely to point to specifics in the Scole Report and say something like "But according to the detailed notes, the medium never moved his hands," or something like that. But we have to remember that, assuming the Scole mediums were using trickery, the authors of the Scole Report were merely witnesses who were taken in by the tricks. Of course their report is likely to, and does, state that they could not have been fooled. This is a perfect example of confirmation bias. These Society for Psychical Research fellows firmly believed they were witnessing genuine spirit phenomena, and desired a positive outcome. They followed the mediums' instructions to the T and acted as an audience only and not as investigators. The Scole Report details the authors' *perceptions* of what happened in the room; no reader has cause to believe it describes what *actually* happened in the room.

Repeatedly, throughout the Scole Report, the authors state that no evidence of fraud or deception was found. For example:

> *There is a further complaint: that we made little mention of the views of people like West or Professor Robert Morris, "who expressed reservations on the basis of their experiences." That is partly because no such reservations were expressed to us at the time... We were looking for evidence of deception... We looked in vain.*

If I go to Penn & Teller's magic show to look for evidence of deception, but I impose the rule that I have to stay in my seat and watch the show as presented, and I'm not allowed to go onstage and examine the performers or the equipment, or watch from behind, or observe the preparations, I guarantee you that I also will find no evidence of deception. Placing illuminated wrist cuffs on the séance mediums, and allowing no further controls, is perfectly analogous to having Teller show you his arms "Hey, look, nothing up my sleeves," then allowing him total control over everything that follows. It can reasonably be argued that the Scole Experiment investigators (whether deliberately or through near-total investigative incompetence) created the conditions of a stage show designed to fool an audience.

The phenomenon most commonly reported in the Scole Experiments were small points of light that flitted about the room, often striking crystals and illuminating them from within, or causing disconnected light bulbs or a small glass dome to light up. Since the mediums banned video gear, there's no way we can really evaluate these claims, other than by reading the Scole Report, which only tells us the perceptions experienced by a few true believers who were present. Mark Edward said these tricks have been commonly performed in séances with laser pointers since the 1970's when they first became available: Strike a light bulb or rock crystal with a laser pointer and it lights right up. An advantage of laser pointers is that the tip can be easily cloaked, obscuring the orifice from anyone whose eyeball is not the target of the beam. We have no evidence that the Scole mediums used such techniques, but their rules also prevented us from establishing that they didn't.

The next most impressive feat was the spontaneous appearance of images on film. During the séance, factory-sealed film cartridges were placed inside a padlocked box. The spirits were then asked to imprint images upon the film. The locked box was then taken and the film developed in the strict constant supervision of the investigators. This feat was repeated many times. One of the investigators, Alan Gauld, wrote critically of how he discovered this locked box could be quickly and easily opened in the dark, which allowed for easy substitution of film rolls. This box was provided by the mediums. Whenever any other sealed container was used, no images ever appeared on the film. Yet even while acknowledging these facts, the authors of the Scole Report still maintain that the film images are most likely evidence of the supernatural.

Perhaps the biggest red flag in the Scole Experiment is the venue in which the sittings took place: a room in the basement of the house in Scole where two of the mediums lived, Robin and Sandra Foy. Rather than controlling the environment, the investigators ceded total control over the room and conditions to the mediums. The séances were held about once a month, which gave the Foys ample time to make any desired alterations to the room. There's no evidence that they did so, but granting them unrestricted opportunity pretty much torpedoed any hope for credibility. The Scole Report states that the room was available for examination before and after every séance, but there's no reason to believe that any truly thorough examination was ever performed; and in any event it's a poor substitute for what the investigators should have done, which was to provide their own room over which the mediums had no control at all. (A few séances were held at other locations, but the Scole Report describes the results from those as "variable".)

The next biggest red flag was the mediums' insistence that the séances be held in complete darkness and their refusal to allow any night-mode video cameras or light enhancement equipment. The mediums' explanation was that they felt such equipment would distract the investigators! That's like telling a pilot that having instruments might distract him from flying. Astoundingly the investigators agreed to this, though they did express dismay, as if their desire and good intentions alone validate their conclusions. Audio recordings only were permitted, but since the claimed phenomena were primarily visual, the audio tapes are of essentially no value.

A third red flag is the fact that there's been no follow-up. If amazing phenomena truly did happen at the Scole Experiment, it would have changed the world. Mainstream psychologists and other academics would have gotten in on it, it would have made worldwide headlines, and it would be repeated in labs everywhere and become mainstream science. They did have the opportunity: skeptical psychologist and author Richard Wiseman sat in on one séance, taking charge of some photographic film, which failed to be imprinted while in his control. But rather than coming away impressed and spreading the word, he summed it up to me in six words: "It was a load of rubbish!"

This same principle explains why we don't see articles from the *Proceedings of the SPR*, like the Scole Report, republished in scientific journals. A scientific investigation of a strange phenomenon assumes the null hypothesis unless the phenomenon can be proven to exist. But the authors of the Scole Report, with complete credulity, did the exact opposite: Their stated position is that the lack of disproof means their séances were real supernatural events. But a primary feature of good research is the elimination of other possible explanations, at which the Scole investigators made no competent effort. Many of the investigators expressed that they were not very convinced by what they witnessed, and it is to the credit of the Scole Report authors that they fairly reported this. But this raises the question: Why then write such a lengthy and credulous report, making such obvious conclusions that these phenomena were real? The lesson to take away from the Scole Experiment is a simple one. Although we all have preconceived notions, we have to put them aside and follow the evidence when we investigate.

References & Further Reading

Keen, M., Ellison, A., Fontana, D. "The Scole Report." *Proceedings of the Society for Psychical Research.* 1 Nov. 1999, Volume 58, Part 220.

Mellenbergh, G.J. *Advising on Research Methods: A consultant's companion.* Rosmalen: Johannes van Kessel, 2008. 143-180.

The Seybert Commission. *Preliminary Report of the Seybert Commission for Investigating Modern Spiritualism.* Philadelphia: J.B. Lippincott Company, 1887.

Troy Taylor. "How to Have a Séance: Tricks of Fraudulent Mediums." *The Haunted Museum.* Dark Haven Entertainment, 1 Jan. 2003. Web. 5 Nov. 2009.
<http://www.prairieghosts.com/seance2.html>

Wiseman, R., Greening, E., Smith, M. "Belief in the paranormal and suggestion in the séance room." *British Journal of Psychology.* 1 Aug. 2003, Volume 94, Issue 3: 285–297.

Wiseman, R., Morris, R. *Guidelines for testing psychic claimants.* Hatfield, UK: University of Hertfordshire Press, 1995.

4. Vaccine Ingredients

Do vaccines really contain the horrifying poisons claimed by antivaccine activists?

In this chapter, we're going to point our skeptical eye at some of the claims made by antivaccine activists, in particular, their lists of frightening chemicals and other dangerous toxins 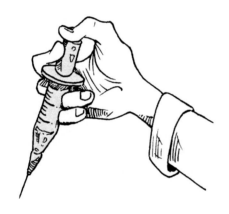 they say are included in vaccines. As it's an important topic and is increasingly in the public eye, we don't want to dismiss these claims out of hand. Rather, we want to have a handy working knowledge of the basics so we're better prepared to deal with such rhetoric when it comes up.

You don't have to go to the antivaccine web sites to find this horrifying list of witch's-brew ingredients. The Centers for Disease Control publishes a detailed list of every additive in every vaccine, sorted both by ingredient and by vaccine. I run my eye down this official list: formaldehyde, aluminum phosphate, ammonium sulfate, bovine extract, thimerosal, amino acids, even monkey kidney tissue. This list is published by the very same government that's assuring us these vaccines are safe. How can that be? Does this mean the antivaccine activists are right, and vaccines are indeed loaded with deadly toxins?

This is a case where cooler heads need to prevail. First, let's start with the premise that every cell of your body is made up of a huge number of chemical compounds, all of which have scary-sounding chemical names. Therefore, we can derive that scary-sounding chemical names, by themselves, are not to be feared. Cooler heads might choose to allow for the possibility that these scary chemicals are added to vaccines because they serve some useful purpose.

When you're exposed to a pathogen, it irritates your body. This irritation is what provokes your immune system to respond, and produce antibodies to fight the pathogen. Vaccines work the same way. They simulate the pathogen in order to produce to right irritation. To prepare your body with the right antibodies to fight some anticipated future pathogen, it's a necessary and expected step for the vaccine to provoke your immune system with a carefully planned challenge. So when you hear antivaxxers charge that vaccines are harmful and irritating, that's quite true, but it's for an important reason and it's very deliberately controlled. This attack on your body to provoke an immunological response is the way vaccines work. It's the way your immune system rolls. You don't strengthen your immune system by eating vitamins or drinking wheatgrass juice or doing yoga or having a coffee enema; you strengthen it by challenging it to respond.

So now that we understand that a vaccine is not pretending to be a shot of Mickey Mouse sunshine, let's take a look at some of these frightening sounding ingredients:

FORMALDEHYDE

Absolutely true. Formaldehyde sounds scary because we see dead animals preserved in jars of it in museums. One of its uses is to sterilize things, and this is why small drops of it are added to some vaccines. Without such sterilization, a vial of vaccine might become contaminated while it's sitting on the shelf. Formaldehyde is used because it's naturally found in the human body, as it's a normal byproduct of digestion and metabolism. When you receive a vaccine shot that was sterilized with formaldehyde, you already have much more of it in your body than you get from the shot. All of this formaldehyde is easily broken down chemically simply because your body is an aqueous environment, and it's harmlessly discharged every day.

Antifreeze

This one is simply untrue. Antifreeze, the poisonous substance used in your car's engine, is ethylene glycol. Because it's so poisonous, antifreeze is not used in food processing or medical equipment, and certainly not in vaccines or other drugs. A less toxic form of antifreeze is propylene glycol, which is not in vaccines either. What is used in some is 2-phenoxyethanol. It's an antibacterial agent used in many vaccines to sterilize them, and also used in wound care as a topical antibacterial. The confusion with antifreeze probably comes from the fact that both are part of the glycol ether family of hydrocarbons, but they are not the same thing.

Mercury

This is the most common claim, and it's the one you've probably heard the most about, so I won't spend much time on it. Some vaccines (but no scheduled childhood vaccines) are preserved with thimerosal, which contains ethylmercury. Elemental mercury is a dangerous neurotoxin, but when it's bound as an organic ethyl, it's easily filtered out of your body by your kidneys and is quickly discharged. This is one reason thimerosal has always been such a safe and popular preservative, and it's still found in many products. Mercury can also be bound as a methyl, which is different, and is much harder for your body to filter out. But fear not; no vaccines or thimerosal ever contained methylmercury, and this scaremongering has no plausible foundation.

Latex Rubber

This one is also completely untrue. Latex is not, in any way, part of any vaccine, and never has been. The source of this claim is the fact that a lot of medical equipment, like syringes and packages, contain latex. Alternatives are always available for people with severe latex allergies. This is a common issue for such people, and has no specific relevance to vaccines whatsoever.

Hydrochloric Acid

Scary sounding, and true. If you pour hydrochloric acid on your skin, you get burned, because your skin is pH balanced. But if you add acid to something that's alkaline, acid brings it back into balance. Hydrochloric acid is used in many industries to bring compounds that are too alkaline to the desired pH level, and the pharmaceutical industry is no different. Some vaccines, once the active ingredients are all added, may be too alkaline; and if injected like that, would cause an adverse reaction. Hydrochloric acid brings the vaccine down to your body's normal pH level of about 7.4. Hydrochloric acid is also the primary digestive acid produced in your stomach, so it's no stranger to the human body.

Aluminum

Aluminum, in various forms, is added to vaccines as an adjuvant. An adjuvant is like a catalyst for the desired irritation, making the challenge even more annoying for your body. It's supposed to be there, on purpose, to make your body react even more strongly. More antibodies are created as a result of the more provocative challenge. Remember: Mickey Mouse sunshine and roses do nothing.

Aluminum is, of course, a neurotoxin, but only at amounts far, far higher than that normally found in our bodies, in the environment, and certainly in vaccines. Just by living and breathing on a planet like Earth where aluminum is the third most abundant element, the average person consumes 3-8mg of aluminum per day, of which less than 1 percent is absorbed into the blood. Vaccine doses are allowed to contain a maximum of .85mg of aluminum; so the maximum dose of aluminum in a vaccine is about the same as the maximum that might get into everyone's blood in a normal day (about what's contained in 33 ounces of infant formula). Most vaccines contain less than this. Studies have proven no difference in neurological condition between children who have had aluminum adjuvated vaccines and those who have not.

Aspartame

Once again: FAIL. Completely untrue. Although any search of the web would have you believe otherwise: The phrase "aspartame in vaccines" is all over Google. So what are these vaccines? I searched the CDC's database of vaccines; nothing. I searched the database of additives; still nothing. I only found only antivax article that mentioned specifically which vaccine aspartame is in, and it claimed only one: The typhoid vaccine Typhim Vi. But it's not true. The additives in Typhim Vi are publicly available and aspartame is not on the list. This is when the antivaxxers are at their worst, when they simply make up lies. This is not constructive for any purpose.

Aborted Fetal Tissue

They sure picked the scariest sounding thing they could think of here! Some viruses don't retain their chemical markers well enough when they're dead in order for the immune system to recognize them, so a very few vaccines are given with the viruses still alive. Growing the weakened viruses means they have to have living cells which they can invade in order to multiply, and these living cells are specific lines that can divide and multiply predictably over a period of many years. Some of these are animal cells, and some are human cells. These cultures are continually reproducing, self-perpetuating lines that are the same generation after generation. The human cells used for this purpose all come from two healthy 3-month-old fetuses aborted in the 1960s by choice. One line, MRC-5, was created in 1966. The other, WI-38, was created in 1962. These two cell lines are used for all the vaccines currently in production worldwide that depend on human cell culture. The cells themselves are not part of the vaccine; just the weakened viruses grown within them.

Human Serum Albumin, or HSA, is a stabilizing protein made from human blood donations. Bovine albumin is also used in a few vaccines. Some vaccines are grown in cultures of monkey or chicken kidney tissue, and when the vaccines are extracted, a few cells from the culture always remain. There's never been any evidence that this might be dangerous. Some vaccines are cultured inside chicken eggs, and some egg protein may remain as a result. This can be a problem for people with severe allergies to egg protein, so these people should avoid these vaccines.

You'll hear all sorts of shock stories about embryonic fluid and cells of exotic animals. Be skeptical of such stories, and you are shocked and concerned, spend five minutes searching the web to find out if that ingredient is actually used; and if so, why; and whether it represents any credible cause for concern. I guarantee you that Jenny McCarthy is neither the first person, nor the best informed, to have considered vaccine safety.

Live Viruses

When a vaccine must use live viruses, formaldehyde is usually used to knock them out, weakening them to the point where they no longer pose a threat, but still alive enough to provoke the desired response. This is not done haphazardly: Finding just the right balance for the vaccine to be effective but not dangerous is hard work.

You'll hear antivaccine activists shout, "Green our vaccines!" What do they mean? Are vaccines environmentally unfriendly? What does "being green" have to do with it? Presumably this is a swipe at vaccine additives that they believe are unsafe or damaging to the environment. Sadly it's too vague of a charge to answer directly. Specific claims can be tested; vague rallying cries cannot. This is where the antivaxxers' movement has taken them: Whenever they've attempted to levy a specific, testable claim, it's easily falsified. Don't let a sound byte as meaningless as "Green our vaccines" carry any clout it has not earned.

Many antivaccine activists believe that a healthy diet is all that's needed to guard against disease. Unfortunately, a healthy diet by itself does not present any immunological challenges. No antibodies are created as a result. Then when a pathogen enters the body, the pathogen wins, and the body becomes diseased. If you focus on your diet or your fitness, but ignore your immune system, expect to look slim and run marathons, but don't expect your immune system to be well prepared should you be unlucky enough to run into polio.

REFERENCES & FURTHER READING

AAP. "Vaccine Ingredients." *Immunization.* American Academy of Pediatrics, 20 Jan. 2011. Web. 25 May. 2011. <http://www.aap.org/immunization/families/ingredients.html>

CDC. "Ingredients of Vaccines - Fact Sheet." *Centers for Disease Control and Prevention.* U. S. Federal Government, 19 May 2009. Web. 11 Feb. 2010. <http://www.cdc.gov/vaccines/vac-gen/additives.htm>

Janeway, C., Travers, P., Walport, M., Shlomchik, M.J. *Immunobiology.* New York: Garland Publishing, 2001. 582-583.

Marshall, Gary. *The Vaccine Handbook.* West Islip: Professional Communications, Inc., 2008.

Ragupathi G., Yeung K., Leung P., Lee M., Lau C., Vickers A., Hood C., Deng G., Cheung N., Cassileth B., Livingston P. "Evaluation of widely consumed botanicals as immunological adjuvants." *Vaccine.* 2 Sep. 2008, 26(37): 4860-4865.

Ribeiro C., Schijns V. "Immunology of vaccine adjuvants." *Methods in Molecular Biology.* 1 Jan. 2010, 626: 1-14.

5. The Baigong Pipes

Do modern metal pipes buried in ancient Chinese stone prove that aliens must have visited?

Should you happen to visit Tibet anytime soon, be sure to stop by the city of Delingha. It's a town of most extraordinary

beauty, nestled on the edge of the Qaidam Basin below a range of Himalayan hills. There you'll find the local residents proudly displaying their most famous distinction. For a few Yuan you can probably get someone to take you to see it. Only a short journey outside of town is said to be a cave, and in this cave

are a series of ancient metal pipes. These pipes predate all known history, and are embedded into the rock itself. They are said to lead through the very mountain, and connect to a nearby salt lake. The explanation? Ruins of a construction project 150,000 years ago, by alien visitors.

The Baigong Pipes are an example of what paranormal enthusiasts refer to as "out of place artifacts", modern objects discovered in ancient surroundings. The Baigong Pipes are described as a sophisticated system of metal pipes, buried in geology in such a way that precludes the possibility of having been installed in modern times. They are located on Mt. Baigong in the Qinghai province of China, about 40 kilometers southwest of Delingha. Most accounts describe a pyramid-shaped outcropping on the mountain, and the cave containing the pipes is on this pyramid. 80 meters from the mouth of this cave is a salt lake (the twin of an adjacent freshwater lake), and more pipes can be found poking up along the shore. Most of the information you can find online about the Baigong Pipes appears to be originally sourced from a 2002 article from the Xinhua News Agency, talking about preparations by a team of scientists about to embark to this remote area to study the pipes. "Nature is harsh here," said one. "There are no residents let alone modern industry in the area, only a few migrant herdsmen to the north of the mountain."

The two lakes are broad, shallow sinks at the low point of the vast Qaidam Basin. Searching for Mt. Baigong is likely to be fruitless: First, the area is largely flat and the nearest mountains are 20 or 30 kilometers away; second, *baigong* is a local word for hill and could mean anything in this context. The southernmost of the two lakes, Toson Hu or Lake Toson, has some low bluffs here and there along its southern and western sides. On its northeastern shore is a point, and it was in a bluff on this point that author Bai Yu once happened to find what he described as a small cave, according to his book *Into the Qaidam*.

Bai was traveling the area in 1996, and described a lifeless lake surrounded by cone-shaped hills. The cave appeared to have been artificially dug, and was triangular, about six meters deep. Nearby were two similar caves, but they had collapsed and could not be entered. But what struck Bai was the array of manufactured metal pipes protruding up through the floor of the cave and embedded within its walls, one 40 cm wide. Following their path outside, Bai discovered more pipes protruding from the surface of the conical hill, and even more of them 80 meters away from the cave along the shore of the lake. Excited, he removed a sample and sent it to the Ministry of Metallurgical Industry. The result was 92% common minerals and metals, and 8% of unknown composition.

Bai proceeded about 70 kilometers to the Delhi branch of China's Purple Mountain Observatory, a high vantage point from where he knew he could get a birds-eye view of the whole region. He saw great expanses of flat, open terrain, and putting two and two together, he concluded that this would make for a fine alien landing site. Unknown minerals and plentiful landing space meant that the Baigong Pipes had to be of alien origin.

Scientists from the China Seismological Bureau visited the lake in 2001 to examine the pipes. Samples brought back to the Beijing Institute of Geology were examined by thermoluminescence dating, a technique that can determine how long it's been since a crystalline mineral was either heated or exposed to sunlight. The result came back that if these were indeed iron pipes that had been smelted, they were made 140-150,000 years ago. Human history in the region only goes back some 30,000 years, and so the alien theory seemed to have been confirmed. The following year the Xinhua news story was published, and the Baigong Pipes entered pop culture as, supposedly, genuine, tangible evidence of alien visitation.

The cave on Lake Toson

If you visit the area today, you'll find a locally built monument to the aliens off the main highway, replete with a mockup metallic satellite dish. Internet forums buzz with the absence of follow-up articles by Xinhua; the natural conclusion is that it turned out the alien explanation was the true one and the Chinese government is suppressing any further reporting.

Cracked.com touts the Baigong Pipes as one of "Six Insane Discoveries that Science Can't Explain".

And although that's where most reporting of the Baigong Pipes stops, it's also where responsible inquiry should begin. When you settle on a paranormal explanation, it means you've decided there is no natural explanation. In fact, when you don't yet know the explanation, you don't yet know the explanation; so you can't reasonably decide that the time is right to stop investigating. But so many do.

Skeptical hypotheses have already been put forward, seeking a natural explanation for the Baigong Pipes that doesn't require the introduction of a wild assumption like alien visitation. The first thing we turn to are geological processes that might explain them. The Chinese have put forth several such hypotheses, including one involving the seepage of iron-rich magma into existing fissures in the rock.

A 2003 article in Xinmin Weekly described how this might work. Fractures caused by the uplift of the Qinghai-Tibet plateau could have left the ground riddled with such fissures, into which the highly pressurized magma driving the uplift would have been forced. Assuming this magma was of the right composition that, when combined with the chemical effects of subsequent geological processes, we might very likely expect to see such rusty iron structures in the local rock. But evidence of this has never surfaced, and the Chinese dismissed this theory. They also noted that the Qaidam oil field would not be able to exist if there were active volcanism in the area as recently as 150,000 years ago.

It was their next theory that ultimately led to a satisfactory explanation, and this theory involved the same hypothesized fissures in the sandstone. But, instead of being filled with iron-rich magma, the fissures could have been washed full of iron-rich sediment during floods. Combined with water and the presence of hydrogen sulfide gas, the sediment could have eventually hardened into the rusty metallic pipe-like structures of iron pyrite found today. This theory was not fantastic, in part because there was no logical reason why the sandstone might happen to be laced with pipe shaped fissures. But the idea of flooding did make sense, given the geological history of the Qaidam Basin.

Three years before Bai Yu took his first peek into the cave at Lake Toson, researchers Mossa and Schumacher wrote in the *Journal of Sedimentary Research* about fossil tree casts in Louisiana. They found cylindrical structures in the soil, thermoluminescence dated from 75-95,000 years ago. The chemical composition of the cylinders varied depending on where and when they formed and in what type of soil. The authors found that these were the fossilized casts of tree roots, formed by pedogenesis (the process by which soil is created) and diagenesis (the lithification of soil into rock through compaction and cementation). The result of this process was to create metallic pipe-like structures, which by comparing the descriptions offered by researchers, appear to be a perfect match for the Baigong Pipes.

The Chinese scientists eventually did come to the same conclusion, according to the Xinmin Weekly article. They used atomic emission spectroscopy to conduct a detailed chemical analysis of the rusty pipe fragments, and found them to contain organic plant matter. Under the microscope they found tree rings, consistently throughout the samples. Once they established that the Baigong Pipes were simply fossilized tree casts, they set about to discover how they got there.

The Qaidam basin was once a vast lake, which has disappeared as the Qinghai-Tibet plateau uplifted the basin to its current elevation of about 2800 meters. Over the millennia, various floods filled the sink with runoff, alluvium, and debris including such fossils. They can now be found wherever such ancient flows deposited them, and it seems that Bai Yu was lucky enough to discover just such a pocket.

And so we end up with a complete story of how rusty iron pipes, tens of thousands of years older than any people who might have forged them, can end up embedded in solid sandstone in such a way as to baffle the average observer. Like many amateur researchers, Bai Yu stumbled upon an extraordinary discovery, but through his lack of applicable knowledge, misinterpreted what he saw. Those who underestimate the Earth's ability to produce fascinating effects are often left to grope for goofy explanations like alien construction projects. I find that the Baigong Pipes are one of the better examples of the folly of stopping at the paranormal explanation, compared to the rich rewards offered by following the scientific method to uncover what's really going on.

References & Further Reading

Beitler, B., Parry, W., Chan, M. "Fingerprints of Fluid Flow: Chemical Diagenetic History of the Jurassic Navajo Sandstone, Southern Utah, U.S.A." *Journal of Sedimentary Research.* 1 Jul. 2005, Volume 75, Number 4: 547-561.

Mossa, J., Schumacher, B. "Fossil tree casts in South Louisiana soils." *Journal of Sedimentary Research.* 1 Jul. 1993, Volume 63, Number 4: 707-713.

Owen, L., Finkel, R., Ma H., Barnard, P. "Late Quaternary landscape evolution in the Kunlun Mountains and Qaidam Basin, Northern Tibet: A framework for examining the links between glaciation, lake level changes and alluvial fan formation." *Quaternary International.* 13 Mar. 2006, Volume 154-155: 73-86.

Xinhuanet. "Chinese Scientists to Head for Suspected ET Relics." *Xinhua News Agency.* 19 Jun. 2002, Newspaper.

Xinmin Weekly. "Alien Ruins Show." *Xinmin Weekly.* 13 Oct. 2003, Newspaper.

Zheng, M. *An introduction to saline lakes on the Qinghai-Tibet plateau.* Boston: Kluwer Academic Publishers, 1997.

6. The Naga Fireballs

What is the source of the glowing balls that rise from the Mekong river each October?

During the full moon every October along the Mekong river between Thailand and Laos, an extraordinary spectacle takes place. A great river serpent winds its way through the darkness, spitting glowing fireballs hundreds of feet into the air. The display is greeted with loud cheers from tens of thousands of spectators cramming the riverbanks. This is the *Phayanak* festival: The welcoming of Lord Buddha as he returns to Earth at the end of the Buddhist Lent, by the great river serpent, the Naga.

According to mythology, the Naga is a gigantic hooded snake. It's prominent throughout Indian and southeastern Asian cultures. It's often believed to be an actual physical animal, but with a supernatural spirit, and many people in the region honestly believe that the animal does live in the local waters. One Naga stands out: The *Phayanak*, or King of the Nagas. Its role varies somewhat among the different cultures, but generally, the Nagas are benevolent servants of Buddha. On the 15th day of the 11th Lunar month, which is a full moon that usually falls in October, the *Phayanak* festival is held. Its center is the Nong Khai province of Thailand, and ground zero, where the fireballs ascend from the river, is the district of Phon Phisai, a small village on the bank of the Mekong river.

Evidence for the existence of the Naga is frequently put forward. There's a photograph widely sold throughout Nong Khai purporting to show a group of about 30 American soldiers holding what appears to be some sort of giant sea serpent. The photograph is titled *Nang Phayanak*, and the caption reads:

The Queen of the Nagas seized by American Army at Mekong River, Laos Military Base on June 27, 1973, with the length of 7.30 meters.

In addition, a Buddhist temple in Nong Khai city, Wat Pho Luang Phra Sai, exhibits some objects that it says are fossilized bones from a Naga, such as a tooth and an egg.

But it is the fireballs themselves that have attracted all the attention, both skeptical and believing. You can see YouTube videos of the fireballs taken during the festival. They are orange specks that streak skyward from way out over the water, rising to a height that's hard to judge but appears to be at least several hundred feet over the course of about three seconds before fading out. As each appears, the crowd reacts like crowds everywhere watching a fireworks show, with appropriate "oohs" and "aahs".

There is both good news and bad news for those who wish to attend the *Phayanak* festival to witness the fireballs. The good news is you are absolutely guaranteed to actually see them with your own eyes. The bad news is that what you'll see are simple fireworks, shot skyward as a tourist attraction. But even though today's fireballs are a harmless festival show, there is no basis for establishing that this is the source of all such fireballs throughout history. Anecdotes persist that the fireballs are still visible at other times of the year and at other locations along the river, and many people say that reports of sightings date back centuries. However this belief that the Naga Fireballs are ancient seems to be merely a locally held understanding; it does not appear to be reliably documented prior to the middle of the 20th century.

In 2002, a television network called iTV sent a crew of investigative journalists to find the source of the fireballs during the festival. On the program titled *Code Cracking*, the team took a boat and snuck quietly up the Laotian side of the river, directly across from Phon Phisai, during the festival. They filmed Laotian soldiers firing tracer rounds into the air, and every time they did, the crowds on the Thailand side of the river reacted with their "oohs" and "aahs". The broadcast was widely perceived as an attack on a sacred belief. Lawsuits and boycotts were threatened against iTV. But, as the saying goes, there's no such thing as bad publicity.

In 2001, an estimated 150,000 people attended the festival. Following the iTV report, and a movie from that same year called *Mekong Full Moon Party,* attendance rose to 400,000 the following year. This brought in 50 to 100 million baht, or as much as 2.5 million dollars, a huge boost for the tiny local economy. The festival has expanded from one day to four days, and the generous Nagas have thoughtfully expanded their fireball performance, now welcoming Buddha on two consecutive nights instead of just one. The wooden seating was replaced with concrete grandstands all up and down Phon Phisai's riverfront in 2003, and the local provincial authorities and the Tourism Authority of Thailand now promote the festival relentlessly. There's even a sign, in English, on the main highway:

> *Welcome to Nong Khai Province on the Bank of the Mekong River - Home of the Naga Dancing Fire Balls.*

The theory that a few palms might be greased to get a few Laotian soldiers to fire off a few rounds should surprise no one.

However, not many people in the west watch Thai television or movies. Virtually any article you read about the Naga Fireballs offers the same "scientific" explanation, which in fact is not very scientific at all. It's just the only one any of these authors have heard, so they go ahead and repeat it. This explanation is familiar to UFO researchers: Swamp gas. It's usually given a scientific-sounding description, having to do with the decomposition of organic matter in the riverbed. This decomposition produces methane gas, which bubbles to the surface. It's a fact that methane can spontaneously ignite when it comes into contact with oxygen (given certain conditions), and the story goes that such bubbles appear, burst into flame, and go shooting up into the sky.

However, there are two fatal flaws with this hypothesis. First, methane can only burn in an oxygen environment within a specific range of concentrations. It can only spontaneously ignite within an even narrower range, and requires the presence of phosphine combined with phosphorous tetrahydride. The needed proportions of these gases are unlikely to be found in nature. Second, in laboratory experiments designed to replicate the conditions needed for spontaneous ignition, the combination of oxygen, methane, and phosphorus compounds burns bright bluish-green with a sudden pop, producing black smoke. Under no conditions does it burn slowly, or red, or rise up in the air as a fireball. So even if the improbable conditions did exist in the Mekong river, the resulting display would not look like the Naga Fireballs.

Nong Khai's main proponent of this natural explanation is a pediatrician, Dr. Manos Kanoksilp, who has made the study of the Naga Fireballs his passion. He believes that the precise conditions require an alignment of the sun, moon, and Earth. He also believes that this particular part of the river is especially high in oxygen (it isn't) and that it's sufficient to spontaneously combust the methane because the sun heats the water to a hot enough temperature (it doesn't).

Whatever is shooting up into the air, you've got to figure that it has some solid mass. When you watch the videos you can see that the red-orange lights go up very fast, consistent with fireworks, small rockets, or even tracer rounds (very much like a 12-gauge shotgun tracer round, which is comparatively slow). How is it possible for any flaming object to move that quickly through the air without blowing out? That's not a problem for something like a firework or a tracer round, things designed for exactly such a purpose. But it's a major problem for a burning ball of gas, which has an insufficient mass to drag ratio to move that quickly through the air. Even a pyrotechnic explosion that billows into the sky rises at a slow rate consistent with hot air rising through cold air; it does not and cannot streak like a bullet at hundreds of feet per second. For the Naga Fireballs to move as they do, they must enclose an object significantly more massive than the air they're moving through. This necessarily means they're heavier than air. And since they're rocketing skyward, this means they must have been physically propelled.

So, from what we can observe, it's actually more plausible that a river dragon is spitting flaming balls of dragon-mucus skyward, than it is for the Naga Fireballs to be naturally produced burning gas bubbles.

Would you hear fireworks or gunfire? I doubt it. The river there is 700 meters wide, or about a half a mile. It takes sound 2.5 seconds to travel that far, by which time all 400,000 spectators are screaming. Combined with the loud music and amplified announcers, you're not hearing anything that someone's not shouting directly into your ear.

The famous photo

And then there's that photograph of the American soldiers holding the Naga at a secret military base. With a little elbow grease, it's possible to track down the actual source of the photograph. As it happens, this picture was first published in a 1996 issue of *Ocean Realm*, by a Scripps Institution of Oceanography scientist at UC San Diego who was one of three called by the US Navy to come and examine a 23-foot oarfish found by a group of SEAL instructors on a beach run at the Naval Special Warfare Center in Coronado, CA. The account was later written up in the April, 1997 issue of *All Hands*, the magazine of the US Navy. The photograph was taken by Lt. DeeDee Van Wormer. This particular oarfish was pretty beat up, and appeared to the Scripps scientists to have met its fate at the business end of a boat propeller.

While sightings of the oarfish are relatively rare, their distribution in salt water is worldwide. Fresh water, like the Mekong river in Laos? Not so much. A Google image search will turn up many such photographs of groups of people holding great long specimens. No mystery here, and no giant sea serpent or military conspiracy needed to explain the photograph on sale in Thailand; and also, *not* evidence of a river serpent.

The lesson to learn from the Naga Fireballs is that, while the historical folk explanation of such stories is almost certainly fictional, the popular "scientific" explanation reported in mass media is often just as wrong. We saw the same thing when we discussed the popular waterspout explanation for frogs and fish falling from the sky; and we see it again here with swamp gas offered as the cause of the Naga Fireballs. When you hear a report of a supernatural phenomenon, the reporter often offers a scientific explanation. It may be right in many cases, but whether it is or not, you should always be skeptical.

References & Further Reading

Carstens, J. "Seals Find Serpent of the Sea." *All Hands: Magazine of the US Navy*. 15 Apr. 1997, Volume 1, Number 960: 20-21.

Cohen, Eric. "The Postmodernization of a Mythical Event: Naga Fireballs on the Mekong River." *Tourism, Culture & Communication*. 1 Jan. 2007, Volume 7: 169-181.

Dow, J. "Mekong Full Moon Party (Movie Review)." *Heroic Cinema*. YesAsia.com, 18 Jan. 2003. Web. 7 Dec. 2009. <http://www.heroic-cinema.com/reviews/mekong>

Gagliardi, Jason. "Behind the Secret of the Naga's Fire." *Time.com*. Time Inc., 17 Nov. 2002. Web. 7 Dec. 2009. <http://www.time.com/time/magazine/article/0,9171,501021125-391567,00.html>

Gampell, J. "Personal Journey: Great Balls of Fire - A Supernatural Serpent Thrills Thailand." *Asian Wall Street Journal*. 9 Nov. 2002, Volume 6, Number 12: 16.

Smith, William Leo. "Oarfish: A Glimpse into the World of the Abyss." *Ocean Realm*. 13 Nov. 1996, Volume 4, Number 2: 28-29.

7. The Antikythera Mechanism

Does this ancient device, 1000 years ahead of its time, prove we were visited by aliens or time travelers?

Imagine the year 1900. The skies are leaden gray, the dark waters all around you shiver with the approach of a distant storm. You're in the warm Ionian Sea in the Mediterranean, sheltering along the coast of the barren and barely populated island of Antikythera, waiting out a storm. It's frustrating because you're on your way home from Africa where you and your crew have been diving for sponges. Your captain, Dimitrios Kondos, thinks the lost time may as well be put to good use, so he orders you below to see what sponges you may be able to find here. You don your copper diving helmet and heavy suit, and they lower you into the depths. Streaks of dull gray light from above shimmer around you as the rocky bottom approaches.

But it's not sponges you find. Minutes later you're back on board the ship, jabbering excitedly, so incoherent that Kondos thinks you have carbon dioxide poisoning. He goes down himself to have a look. And what Kondos and his crew bring up over the subsequent two years comprises one of archaeology's great finds, one that truly challenged our understanding of the history of technology.

Chief among the finds was what has become known as the Antikythera Mechanism, fragile chunks of green corroded bronze, that when picked apart, revealed unexpected mechanical components, mainly gears. The device was surprisingly complex. At first it was thought to be a clock, but when Greek inscriptions were found, it turned out to be a sort of astrolabe for predicting eclipses and moon phases and the positions of the planets, of unprecedented sophistication. So sophisticated, in fact, that everything we knew told us that the Antikythera Mechanism was a full 1,000 years out of place.

The shipwreck, known as the Antikythera Wreck, has been dated to the first century BCE. The Antikythera Mechanism dates from the century before that. And then, so the popular version of the story goes, nobody on Earth had either the astronomical knowledge, or the mechanical know-how, to construct such a device until a millennium later. Some have said the Antikythera Mechanism is therefore proof of time travel, alien visitation, or Atlantis.

Physically, the device was about the size of a shoebox, with wooden sides and bronze faces. On the front face were two large and three small output dials. On the back were three concentric output dials. To operate the device, you turned a crank on the side, which rotated at least 30 gears inside the machine, some of which were epicyclic. The hands that went round each of the two large dials swept over spiral slots, with a pin on the arm that rode in the slot, similar to a needle following the groove on a record. By setting some preferences, such as what type of calendar you wanted to use, and turning the side crank to select the current date, you could learn all sorts of things: Whether this was an Olympics year, when the next solar and lunar eclipses were (by date and hour), where the twelve constellations

A portion of the mechanism on display in Greece

were along the ecliptic, the phase of the moon, and the positions of the five planets known at the time.

Although we now know what the device did, we're not sure what its use was. By its construction in bronze, which readily corrodes, we know it was not designed for navigation at sea. Astronomers and astrologers probably could not have afforded it. It could have been used as an education tool. Most likely it was built for wealthy Romans who had some interest in its features, probably not too different from early adopters who wanted to have the first iPhone with all the cool apps. The wreck was laden with other objects of great value, most notably a vast hoard of coins and a Peloponnesian bronze sculpture, a larger-than-life young man called the *Ephebe*.

So what about these claims that the mechanism is 1,000 years out of place, and no humans had the knowledge to make something like it? Does this prove that aliens, Atlanteans, or time travelers must have been involved? It is a fact that the Antikythera Mechanism is substantially more complicated than any other mechanical devices known from its time. Specifically, it's one of the earliest known uses of meshing gears. But contrary to the popular telling, it's not the oldest. Gears were used to drive doors and lift water in India as early as 2600 BCE, two and a half millennia before the Antikythera Mechanism. Aristotle described the function of gears in the 4th century BCE. 100 years later, Dionysius of Alexandria used gears in his automatic arrow firing machine gun. The Greek National Museum contains examples of epicyclic gears from the period. Archimedes was making all sorts of mathematical and mechanical inventions at the time. For hundreds of years, Greek astronomers had been studying the movements of heavenly bodies, and by Archimedes' lifetime, all the motions replicated on the Antikythera Mechanism were known to science.

We know a lot about where and when the device was made from the inscriptions in the bronze. The back face is covered with instructions for its use, as is its inside if you open the device. These include descriptions of the controls, various calendars, and mentions of the celestial bodies tracked by the device. By the language and terminology used, as well as by the context of its find among the other artifacts recovered from the Antikythera Wreck, we now have a pretty good idea of where and when it was built: The middle of the 2nd century BCE, probably in Syracuse or Corinth.

What we don't know is who built it, but there are some good candidates. It is assumed that Archimedes, who died several decades before the device was built, left behind a tradition of scientists who continued his work and built upon his inventions, and the device could have come from this school. Another leading contender is Hipparchos, perhaps the greatest of early astronomers, who was in his heyday when the Antikythera Mechanism was constructed. Most notably, Hipparchos was the first to devise a mathematical model to predict the anomalies of the moon's movement, and the Antikythera Mechanism contains a gear set to reproduce exactly these computations.

Most archaeologists agree that this particular device was neither unique nor the first of its kind. Two factors contribute to this: First, its design is quite refined, which is not consistent with a prototype. Second, an object as expensive and complex as this would typically be made in a series in order to recover the costs of design. Why, then, are its siblings not found? Probably because they were made of bronze, and bronze was highly recyclable and valuable. Few commonplace bronze objects from the ancient world survive for this reason, except for those that were lost at sea and thus escaped recycling. If there were other computational devices made in the period, it is not surprising to archaeologists that they were lost to history and are unknown.

The surviving fragments of the device at the Greek National Archaeological Museum are too fragile to travel, and so in 2005, two teams brought their equipment to Athens to perform the most advanced imaging studies to date. Hewlett Packard's team performed polynomial texture mapping to high resolution images of the inscriptions, made with lighting from all different angles to reveal every possible bit of detail. A company called X-Tek Systems brought their 8-ton x-ray machine called the BladeRunner to the museum, all the way from the UK, and made CT scans. CT, or computed tomography, is the process of creating a 3-D image from slices; in this case, slices from x-ray images. We now have extremely detailed maps of the internal mechanism and transcripts of all the surviving inscriptions.

And, as a result, we now know that the ancient Greeks were building far more advanced computational devices than we used to give them credit for. We knew they had the knowledge, we just didn't know they were translating it into bronze so exquisitely.

Antiscience people love to point to cases like the Antikythera Mechanism as examples of science being wrong. They gloat over their belief that historians have been embarrassed, careers shattered, books proven to be in error. They imagine that researchers at universities everywhere are being fired or stripped of their credentials. They believe this case adds to an ever-growing mountain of proof that science is, itself, destined to inevitable failure, and that *enlightened* scientists should abandon their practice and turn to faith in the supernatural.

In fact, nothing could be further from the truth. A find like this, that substantially revises our understanding of history, can be the crowning achievement for a scientist's career. Contrary to what antiscience would have you believe, scientists do not fear new discoveries, they long for them. Every action dictated by the scientific method seeks to learn something new, to revise and improve our understanding of nature or history. Major finds represent major improvements to our theories. Thanks to the Antikythera Mechanism, we now have a better understanding of technology in the ancient world, and new directions for researchers to turn. The idea that a discovery like this is embarrassing, or exposes weakness in the scientific method, is absurdly upside-down and backwards.

Many times I've heard the argument made that scientists fear new discoveries because it would threaten their grant money, so it's in scientists' best interests to cover up anything new. All you have to do is look at what's being funded these days to see how wrong this particular conspiracy theory is. A glance at the National Science Foundation's list of recent funding awards teaches one lesson: If you want grant money, be a maverick, have something new and exciting. What's a financial incentive for scientists to look for discoveries that challenge our worldview? New discoveries attract grant money, and grant money leads to more new discoveries. There is no money in continuing to grind over what we already know. So please, can we put this particular conspiracy theory to bed?

It's not every year that we find something so historically significant, and that we learn so much from. The rewards we gain from increasing our knowledge by studying them are inestimable. Those who dismiss such finds as alien, or otherwise not part of history, miss out on that knowledge. The lesson to learn is that when you're confronted by a discovery, stopping at the popular supernatural explanation is guaranteed to lead you nowhere. Instead, you should do what science suggests, and be skeptical.

References & Further Reading

Edmunds, M. "Project Overview." *The Antikythera Mechanism Research Project.* School of Physics and Astronomy, Cardiff University, 1 Jan. 2007. Web. 26 Jan. 2010. <http://www.antikythera-mechanism.gr>

Fine, J. *Lost on the Ocean Floor: Diving the World's Ghost Ships.* Anapolis: Naval Institute Press, 2005. 83-85.

Freeth, T. "Decoding an Ancient Computer." *Scientific American.* 1 Dec. 2009, Volume 301, Number 6: 70-83.

Hardersen, P. "Mickey Mouse Discovers the 'Real' Atlantis." *Skeptical Inquirer.* 1 Jan. 2002, Volume 26, Number 1: 42-43.

Kanas, N. *Star Maps: History, Artistry, and Cartography.* Chichester: Praxis Publishing Ltd., 2007. 240-241.

Price, D. "An Ancient Greek Computer." *Scientific American.* 1 Jun. 1979, Volume 200, Number 6: 60-67.

8. Is Barefoot Better?

Some advocate that going barefoot is better for the health and strength of your feet.

In this chapter, we're going to let our dreadlocks down and take a rational, science-based perspective on a trend that seems, at face value, like just another nonsense hippie claim from the "anything natural is good, anything modern is immoral" crowd: The idea that we'd all be better off being barefoot. Whether you run marathons or give boardroom presentations, barefoot advocates claim that barefoot, the way we evolved to walk and run, relieves and prevents orthopedic injuries.

I'll freely confess that the first time I heard this claim I scoffed, it bears so many of the red flags of pseudoscience. These red flags include the implication of a medical/industrial conspiracy to keep us injured by selling us expensive shoes and orthopedic treatments, and of course the ever-present all-natural fallacy. But what grabbed my attention was that I also noted that products like expensive running shoes also sport a major red flag that I'm keenly aware of. Until extremely recently in the history of our species, nobody had expensive running shoes, or even any

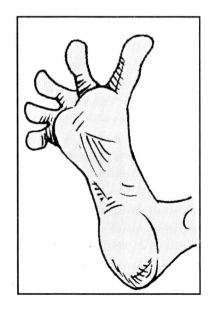

shoes at all, and we did just fine. I'm reminded of how I always smirk when I see new parents paying double for special baby apple juice in the supermarket: Even today, most babies in the world are lucky to get a twig and a dirt clod to eat each day, yet they grow up fine. Do I, the product of millions of years of evolution creating a bipedal animal optimized for walking and running barefoot on the savannah, really need a $100 pair of super-duper running shoes to make my feet work properly? My default answer for such questions, based on evolution, is "probably not". So I believe this barefoot question does deserve a really close look.

Logically, the argument in favor of going barefoot — both for everyday locomotion and for athletes — is a compelling one. It is a fact that that's how our feet evolved to work. Record setting runners like Zola Budd and Abebe Bikila have proven that barefooters are perfectly capable of competing at the highest levels of the sport, no shoes needed. A friend of mine visited Africa and brought back pictures of the feet of some guy who never wore shoes his whole life. They didn't much look like human feet, more like great big thick gray leather pads. He could probably walk on any kind of surface at any temperature without the slightest inconvenience. And forget worrying about arch support; he had no discernible arches. If you don't mind looking like a hobbit, his way of life might not be so bad. Go anywhere, anytime, with no concerns and no stability problems. After all, that's exactly how the overwhelming majority of Homo sapiens have lived their entire lives, throughout our entire 500,000 year history, and how some do still today.

There are quite a few web sites advocating the barefoot way of life (like Barefooters.org and RunningBarefoot.org). Most of the benefits they tout have to do with freedom, comfort, and the carefree lifestyle. But they also cite some health benefits. One 1949 study published in *The Journal of the National Association of Chiropodists* found that bare feet avoid most risks of foot fungus. Over 100 Chinese rickshaw coolies were interviewed, all of whom trotted all day every day on hard pavement pulling a rickshaw with bare feet, and aside from some temporary pain and swelling when they began their careers, none reported any foot problems. The study concluded "Shoes are not necessary for healthy feet and are the cause of most foot troubles," though it does appear that this conclusion is probably premature, based on this data alone. Ill-fitting and restrictive footgear was claimed to be particularly at fault, but since modern Nikes are not best described as "ill-fitting and restrictive", it may be hard to translate the 1949 conclusion to today's runners.

The principal medical claim put forward by the barefoot proponents is that wearing shoes weakens the muscles in your foot, through disuse. While this sounds like it should be obviously true, it does not appear to have been thoroughly studied or proven. And there's a good argument that the opposite might be true. When you wear a shoe with a sole of average thickness, around 22mm, the lateral leverage angle on your ankle is significantly increased, which, logically, should make it more prone to injury; and requires greater strength in the muscles and tendons that stabilize the ankle to compensate.

A search of the literature reveals that most researchers complain of a lack of good studies comparing injury incidence between barefoot runners and shod runners, so it's premature to make any kind of authoritative statement that either barefoot running, or running shoes, help prevent injuries. If you hear this claim made by either side, that claim is not yet supported by a good body of research. But what has been established pretty clearly is that running barefoot keeps you more up on your toes, closer to the stride of animals, while shoes let you strike harder with your heel. This means (and nobody really disputes this) that shoes make you run in an unnatural way; or at least with a stride that's notably different from that which our ancestors evolved.

This heel-to-toe stride encouraged by shoes does appear to be correlated with pronation, but the link is not necessarily causal. It's easy to imagine how constant landing on the heel could cause cumulative stretching of the ligaments and eventually roll the foot inward toward the arch. However, whether this cause exists or not, it's not clear that such pronation is a problem. There doesn't seem to be any proven correlation between pronation and injury incidence. But the claim that feet are weakened by frequent shoe wearing are everywhere, and are easily believed, since it sounds so logical.

In response, shoe manufacturers have been quietly entering the arena, making less shoe-like shoes. The Vibram FiveFingers is basically a glove for your foot, providing little more than a thin sole to protect you from sharp objects. Vibram says "Stimulating the muscles in your feet and lower legs will not only make you stronger and healthier, it improves your balance, agility and proprioception."

The Vivo Barefoot is slightly more conventional looking but has no rigid structure and only a thin puncture-proof sole. The benefits Vivo claims for going barefoot is that it "Strengthens the muscles in your feet; realigns your natural posture; feeling the ground stimulates sensory perception; and flexes your feet as nature designed."

All of these claims, I think, have to be regarded as marketing messages. They are not the result of proven research. They sound satisfyingly logical, and some of them may well be true; but at this point, we don't know that they are. Yes, we did evolve wearing bare feet, but that doesn't necessarily mean it's the best thing to do.

That evolution has made us into bipeds that walk and run barefoot across the savannah is not a perfect argument that we're well adapted to do so. There are many examples in nature of creatures that evolved detrimental traits. The giraffe's laryngeal nerve runs all the way down its neck into its chest, loops around its aorta, then runs all the way back up to its larynx; making it absurdly long and prone to many types of failure. The Irish elk developed antlers so large that the energy required to grow them exceeded the available food sources and the species became extinct. The mating ritual of the Kakapo flightless parrot is more likely to attract a predator than a mate. The retinas in all vertebrate eyeballs are inside out, creating an unnecessary blind spot. The list goes on forever. The point is that evolution does not create perfectly adapted creatures; it creates adequate creatures.

The human body too is full of evolved points of failure, and we've learned to fix many of them. If shoes do indeed improve our ability to walk, they would be just one of many examples where we've used medical science to improve our bodies' inherent weaknesses. Our appendix doesn't seem to serve much purpose except to potentially kill us, so we've developed the appendectomy. Wisdom teeth try to force too many teeth into too small of a jaw, so we routinely remove them. Nearly everyone has bad eyes, so we correct our vision. Orthopedically, our knees' lateral retinaculum exacerbates many knee problems, so some modern knee surgeries include a routine severing of this structure.

There is other anatomical evidence that the human transition to bipedality is not a very complete one. Walking upright has left human females with a pelvis that, relative to other primates, doesn't allow much room for the birth canal. This makes giving birth more dangerous for humans than it is for the great apes that knuckle walk. And then there's this strange foot we have. While nearly every other animal on the planet walks on its fingers and toes, we lay one additional segment down and walk on our ankles, and even crawl on our wrists as babies. Trading one joint for extra surface area was fine for gripping tree branches, but it makes little sense for walking on flat ground. Among other things, it leaves us with one fewer joint to absorb twists, making us more prone to ankle and knee injuries.

So is there a clear, science-based verdict on the barefoot issue? From my research, I conclude that the jury is out; but any benefit that may be found by either wearing shoes or going barefoot is likely to be small, and to differ widely among different people, based on their habits and their anatomy. The most interesting revelation, for me, was that barefoot running does indeed appear to be a perfectly viable option for both athletes and casual joggers. The only real risk is puncture injury, which you can solve by either developing feet like that African guy, or by getting some of those barefoot-wannabe shoes. Once you get past some initial pain and swelling, similar to what the rickshaw coolies went through, you're probably no more or less likely to develop any significant injuries than if you were to wear shoes.

Zola Budd, by the way, is long retired but still runs tens miles per day in her native South Africa. But things have changed for her. "I no longer run barefoot," she told the UK *Guardian* in 2005. "As I got older I had injuries to my hamstring. I found that wearing shoes gives me more support and protection from injuries." But hers is only one data point, and is anecdotal. We just need a few thousand controlled data points before science can make such a declaration.

REFERENCES & FURTHER READING

Editors. "Barefoot Running FAQs." *Vibram FiveFingers*. Vibram USA, 1 Jan. 2010. Web. 25 May. 2011.
<http://www.vibramfivefingers.com/faq/barefoot_running_faq.htm>

Harmon, K. "Running barefoot is better, researchers find." *Scientific American Observations*. Scientific American, 27 Jan. 2010. Web. 2 Feb. 2010.
<http://www.scientificamerican.com/blog/post.cfm?id=running-barefoot-is-better-research-2010-01-27>

Kerrigan, D., Franz, J., Keenan, G., Dicharry, J., Croce, U., Wilder, R. "The Effect of Running Shoes on Lower Extremity Joint Torques." *Physical Medicine and Rehabilitation*. 1 Dec. 2009, Volume 1, Issue 12: 1058-1063.

Moen, R., Pastor, J., Cohen, Y. "Antler growth and extinction of Irish elk." *Evolutionary Ecology Research*. 1 Feb. 1999, Volume 1, Number 2: 235-249.

Olshansky J., Carnes B., Butler R. "If humans were built to last." *Scientific American*. 1 Mar. 2001, Volume 284, Number 3: 50-55.

Ridley, M. *Evolution*. Oxford: Wiley-Blackwell, 2004. 282.

9. Emergency Handbook: What to Do When a Friend Loves Woo

How you can help a friend or loved one with a potentially harmful pseudoscientific belief

It's the #1 most common question I get: My wife, my friend, my mom, my boss, is investing their health or their money in some magical or fraudulent product/scheme/belief. What can I do about it?

This is a tough situation to be in. Whether it's a loved one who's ill and is being taken advantage of by a charlatan selling a magical cure with no hope of treating the illness, or a friend who's out of work and is going into deeper debt to buy into a hopeless multilevel marketing plan, it's really hard to watch. The hardest is when they have a real problem and are expending their limited resources trying to solve it with a medieval, magic-based system that you know can't possibly help. But all too often, they think it's helping. Cognitive biases, anecdotal thinking, placebo effects and cognitive dissonance combine to build a powerful illusion that our brains are hardwired to believe in. At some point, it falls to a caring friend to try and rescue them with a candle of reason.

You're up against a foe who's far more formidable than you might think. This isn't like settling a bet with a friend where you can look up the answer on Wikipedia, see who's right, then buy each other a beer. You're going after someone's religion. You're setting out to talk someone out of believing something that they know to be true, for a fact, from their personal experience. That right there makes your task nearly impossible, but it's worse. Their belief has spiritual underpinnings that make it deeply moral and virtuous. Imagine if someone came to you and flashed a magazine article that said it's best to turn your children out into the street and never talk to them again. It's not only unconvincing, it's laughable. Your effort to talk someone out of their belief in their sacred cow is likely to be just as laughable.

So what should you do, give up? You may be surprised to hear it from me, but I advise you to do just that, in many cases. Know which battles to fight. Weigh the risks. Consider the context of your friend's belief: Is he in imminent danger of harming himself or others? Probably not; and if not, this may not be the time to take what might be your only shot. So I want to make this a rule: Before you decide what to do, consider the risks and the context. How terrible are the consequences of your friend's belief? Think that through comprehensively. Make sure you have a good understanding of the risks to your friend if you do nothing, and the risks to your relationship if you attack their beliefs and (in all probability) fail to convince them. It may well be that this first strategy I'm going to present is the safest.

Strategy #1: Do Nothing

Doing nothing now doesn't mean giving up. When you choose not to confront your friend's current weird belief, there's still an effective strategy for helping him out that you can follow. By accepting and tolerating your friend's weird belief, you're actually setting yourself up to be in a position of great influence the next time something weird comes down the line. Your friend likely knows that you're a skeptical person, and eventually he'll recognize that you've been putting up with his weird belief and saying nothing. In fact he may someday ask you, "Hey, you know I believe in this weird thing, how come Mr. Cynical Skeptic has never tried to talk me out of it?"

Ask "Is it important to you?"

"Yes."

"You're important to me."

Think what a powerful message that sends. It may sound corny, but it's a statement that your friend will always remember. You've just communicated that your friendship is more important than your "evil debunking hobby". You've made it clear, unequivocally, that you don't want such differences to come between you.

And now look at the position you're in. You're trusted. You're an ally at the most important and fundamental level. This is exactly where you need to be if you want to be influential on someone. You can now begin to introduce critical thinking using topics that are more about exploration than confrontation, and this is a journey you should take together. Next time you're in the car together, play a few Skeptoid podcast episodes. Play episodes like *The Baigong Pipes*, *Is He Real or Is He Fictional*, *The Missing Cosmonauts*, and *When People Talk Backwards*. Topics such as these do not attack or challenge anyone, they instill an appreciation and a passion for the value of critical thinking. Once introduced, I find that most people want more.

Gather every bit of skeptical material you can find that you know will interest your friend, and that does not attack or challenge his belief. So long as you remain a trustworthy friend and not an irrational adversary, you're in a position to introduce him to the fundamentals of critical thinking, and to the value and tangible rewards of reality. Don't underestimate the value of seeds that are well planted in a good environment. If your friend comes around on his own, his growth is far more complete than any that's forced upon him.

Always remember the story of the little boy who couldn't get his pet turtle to come out of its shell. He tried to pull on its head, he shook it, he squirted water, he did everything he could think of. But the turtle wouldn't come out. Then his grandfather took the turtle and placed it on the warm hearth, and within a minute the turtle was out of his shell. The little boy never forgot that lesson.

Strategy #2: The Intervention

Sometimes the situation is urgent and you don't have time to do things the easy way. There might be a medical crisis, an emotional crisis, or a financial crisis, and an immediate intervention is needed. Sometimes a friend's situation is dire enough that helping him is worth the loss of the personal relationship. In these cases, and probably only in these cases, would I suggest a confrontational approach. And to do this effectively, draw on the established principals of the counseling intervention.

First you want to gather a group of friends or family, and you need to meet with them separately. Try to get a group, but even if there are only two of you, it's worlds better than just you by yourself. Your next task is to present your evidence to the group that the magical system your friend is relying on is pseudoscientific and cannot help him. Do not expect them to accept what you say at face value, and do expect that some of them might buy into the magical system as well. Be prepared. Show your work. Print out pages from the web. Use the *Science Based Medicine* blog, use *Skeptoid*, use *Quackwatch*, use *Swift*. Search the best sources and have all your ducks in a row. The most important thing you need to do at this stage is to be certain that everyone in the group is united in their understanding of the useless, pseudoscientific nature of the magical sacred cow.

Tell the group why you're concerned about your friend and why the help is urgently needed. Be prepared to explain why you feel an intervention is warranted. And this is important: Don't merely be prepared to show that the magical sacred cow is useless, you must also have an alternative path — one that is proven to provide the kind of help needed — to suggest to your friend. Make sure everyone's in agreement that an intervention is warranted, and that a better alternative path is needed. If they're not, only invite those who are to proceed.

The main criticism of counseling interventions is that they are ambushes. Not only is it just plain wrong to ambush someone, it creates the practical problem of putting your friend on the defensive. So I don't propose making it an ambush. My recommendation, which you may or may not choose to follow, is to call your friend up and say "Hey, Jim-Bob and Bubba and Sally-Sue and I want to come over and talk to you about your cancer," or your new business, or your psychic friend, or whatever the problem is.

Now, of course, conducting the intervention is up to you. I feel that trusted friends who can speak knowledgeably about the subject carry more weight than showing the printed-out articles from the web, but leave them for your friend to read. Anyway, it's going to be a really crappy hour, it's not going to be fun for anyone, but with some luck you may just make a big difference in your friend's life. He may not love you for it, but the idea's to help him, not to win brownie points for yourself.

Strategy #3: Be There

In some cases, doing nothing may seem too slow, and an intervention may be too harsh and unwarranted. In these situations I often recommend that you just "be there" for your friend. Your skeptical cat is probably already out of the bag to some degree, so your friend's radar is probably already up just waiting for you to launch into him about his sacred cow.

What you may have is an awkward imbalance of a close personal relationship and an ideological divide. This situation gets thrown on me all the time, when I meet someone, or someone introduces me and has told them what I do, I'll sometimes get "Oh, you're that cynical person I've heard about." This is, of course, both wrong and insulting. I've gotten this so many times that I've learned to just swallow it. But nevertheless, the awkward divide exists, and the best way to handle an awkward situation is to openly acknowledge it.

The wording here is to difficult to get right, but at an opportune moment, you might want to say something like "Hey, I know you're really into your thing, you know I'm really into consumer protection, so we have a disagreement. If you ever want to talk about it, I'm happy to; but I don't want it to be a problem, and I'm fine with just acknowledging we have a disagreement and leaving it at that." The wording that's hard is declaring your own position. Consumer protection, critical thinking, things that are proven; you want to make your point but you don't want to choose weasel words that sound insulting.

From that point, you can follow the Do Nothing strategy and introduce articles that you know you'll both appreciate. This method just fast tracks it somewhat, in that the door is wide open to discuss your friend's particular sacred cow at any time. But unless there's an imminent risk of harm, I tend to always let the friend bring it up, and I never try to drive the wedge or create a conflict myself.

It's perhaps ironic that those of us who want to provide *actual* help, instead of magical or imaginary help, are usually considered the bad guys, and we're the ones who have to tread lightly. But that's the reality of the situation, and we should take extra care to insure that our influence is a positive one.

References & Further Reading

Ainsworth, P. *Understanding Depression.* Jackson: University Press of Mississippi, 2000. 117-120.

Bowen, D., Strickler, S. *A Good Friend For Bad Times.* Minneapolis: Augsburg Fortress, 2004. 34-67.

Driscoll, W., Kempf, D., Broda-Bahm, K. *Argument and Audience: Presenting Debates in Public Settings.* Westborough: IDEA, 2004. 185-186.

Golanty, E., Edlin, G. *Health and Wellness.* Sudbury: Jones & Bartlett Learning, 2007. 71-73.

Hammer, E., Dunn, D., Weiten, W. *Psychology Applied to Modern Life: Adjustment in the 21st Century.* Florence: Cengage Learning, 2008. 250-251.

Johnson Institute. *Training Families To Do a Successful Intervention.* Minneapolis: Johnson Institute-QVS, Inc., 1996. 14-42.

Ketcham, K. *Teens Under the Influence: The Truth About Kids, Alcohol, and Other Drugs.* New York: Random House, Inc., 2008. 17-19.

10. Martial Arts Magic

Some call it Bullshido: Martial arts tricks like touchless attacks and the Touch of Death.

In dojos all around the world, martial arts masters practice mysterious forms of attack. They can kill or render an attacker unconscious with a single touch, or sometimes, with no touch at all. The *dim mak* and *kyusho jitsu* are just some of the secret techniques reserved only for the masters, that are jealously guarded, and will not be taught to just anyone. Some call these techniques *bullshido*.

Bullshido is, obviously, a joke term which mocks made-up or exaggerated martial arts claims. *Bullshido* comes in many forms. The touch of death and the knockout without touching are just a few of the most popular, originally made famous by the stories telling this is how Bruce Lee was killed (in fact he died of cerebral edema after a dinner party, possibly due to a drug interaction). *Bullshido* also encompasses newly invented martial arts techniques by self-described masters who market themselves as the founders; schools claiming to be too exclusive to let just anyone in (sometimes called McDojos); and claims by instructors of having been taught by various great masters, the missing documentation of which is sometimes explained as being sacred or hidden away in a remote Asian temple.

The many various forms of *bullshido* have long been criticized by legitimate martial arts practitioners, and dismissed merely as marketing claims intended to attract students to a particular school where one of these supposed masters teaches. *Bullshido* practitioners shoot back that such naysayers are merely crying sour grapes because they have not yet learned the secret techniques, or achieved the special level.

The most famous of example of *bullshido*, which you've no doubt seen several times over the past couple of years, involves instructors who claim to have developed a technique of rendering an attacker senseless without actually touching him. The volunteer attackers are always the instructor's students in these videos. They'll charge at him one after the other, and as he punches or swipes at the air, they'll often dramatically fly back as if struck by a train. Every time an outsider volunteers to receive the touchless attack, the instructor either fails with some excuse, or refuses on the grounds that it would be too dangerous.

Harry Cameron is a martial arts instructor who goes by the moniker "The Human Stun Gun". Danielle Serino, a reporter on Fox Chicago's prime time news, decided to check out his claim on her segment *Does It Work, Danielle?* She watched him knock out some of his students by, basically, what amounted to little more than going up to them and shouting Boo! Danielle got suited up and volunteered to have the Human Stun Gun knock her out without touching her. He refused, saying it would be too dangerous for her, even though she went to the trouble and expense of having a team of paramedics standing by. However he was willing to actually punch her on the side of the head. Even that didn't have any real effect except to tick her off.

Danielle decided to give him the opportunity to prove his ability on someone he wouldn't be afraid of hurting, namely, a group of *jiu-jitsu* athletes from another gym who were not his students. His touchless attacks had no effect on any of them. Predictably, he had an explanation handy: Natural athletes like these students learn to "translate the energy" and are not affected by it. I guess Cameron's own students are not as enlightened. One red flag waving over Cameron's head is that he says he was instructed by George Dillman, often cited as one of the great pillars of *bullshido*.

There's also a famous YouTube video you may have seen where an elderly martial arts master, Kiai Master Ryukerin, does the same thing to a room full of his students, easily sending them all tumbling with waves of his hand. He offered $5000 to any modern Mixed Martial Arts athlete who could beat him. One guy took him up on it, and in front of Japanese TV cameras, casually beat the poor old guy to a pulp. It's actually a little sad, and hard to watch. Did Ryukerin actually believe that he had this power? Was it a mass delusion shared between him and his students, or was it all part of the show, and Ryukerin hoped that his actual martial arts skills would defeat the MMA guy? The only thing we know for sure is that his touchless attack failed.

But history, especially recent history, is full of people who claim to have this ability, and are happy to demonstrate it so long as the conditions are under their own control. A Russian martial art called *Systema SpetsNaz* claims the same touchless knockout, as do most others. An American who called himself Count Dante was one of the prototypical *bullshido* practitioners, and spawned an entire subculture through ads in the back of comic books offering the "World's Deadliest Fighting Secrets".

The touchless attack is universally claimed to work by disrupting the victim's *qi*. *Qi* is a hypothetical energy field, the body's life force, which flows through hypothetical channels called meridians and was first postulated in prescientific ancient times to explain why people are alive and inanimate objects are not. *Qi* has no describable or detectable properties, and the only evidence that it might exist is the anecdotal claims of believers who say they can sense it. Yet even they cannot describe it, and cannot detect it under controlled conditions. So while we can't state that *qi* does not exist, we can state that its existence has not been demonstrated.

So if we can't prove that a touchless attack is real, what about the *dim mak*, the so-called "touch of death"? There are two varieties of this. The first is a single sharp blow, and the second is a simple touch or series of touches, as often dramatized in movies.

The single sharp blow is well established to be real, and I'm not even talking about things like a piano falling on you. People can absolutely be killed with a single blow. It's rare and it requires just the right circumstances, but it happens. One way is *commotio cordis*, which is a blunt force blow to the chest, which need not be severe enough to cause any physiological damage; but if it happens just right it can stop the heart. It's usually seen with baseballs, hockey pucks, bullet strikes to bulletproof vests, or even martial arts blows. No *bullshido* here: Punch someone in the heart just right and you can actually kill them.

The carotid artery in your neck is another vulnerable spot, but it requires damage to the artery that causes a stroke. There are many hoax videos on the Internet showing a master knocking out a student with a quick, light chop to the side of the neck. These are sometimes rationalized with the claim that even such an extremely brief interruption to the blood flow to the brain produces unconsciousness. This is false. If it were true, you could do it to yourself. Due to the potential for lethal damage to the artery walls, I don't recommend experimentation.

In boxing they speak of punching someone on the temple as "the button": Hit someone just so, on the button, for an immediate knockout. This is simply a concussion caused by a sudden shock to the head, crushing the brain against the inside of the skull. Contrary to popular belief, the temple is no better or worse a target for concussion than any other point on the skull (except the jaw, which can move and thus absorbs part of the energy of the blow). Aside from concussion, the temple is a dangerous place to be struck, but not because there are any special nerves there. The skull at that point is quite thin and fragile, and right under it is the middle meningeal artery. If it's ruptured, the hemorrhaging is very dangerous, and quite likely fatal. No touch or blow to the temple that does not break the skull or cause a concussion is likely to be especially harmful.

The ability to disable, paralyze, kill, or render a person unconscious with touches or a series of touches to special nerve points on the body, often referred to as acupuncture or acupressure points, is also completely fictitious. Nerves do not serve this function. As has been proven time and time again in clinical trials, acupuncture using traditional acupuncture points produces results no better than random points on the body. By any reasonable analysis, this means acupuncture points, as they are traditionally defined, do not exist. There is nothing special or unusual about the nerves or other anatomical features at so-called acupuncture points, and no clinical effect can be demonstrated by using them. Thus, by extension, any martial arts attack that claims acupuncture points as its foundation is based on a false premise.

There are certainly spots on the human body that are vulnerable to injury or that produce sharp pain if struck. The family jewels are one obvious such place. The nose is another. There are various techniques for damaging knee or elbow ligaments with relatively little effort. Various spots on the body produce involuntary reflexes if struck properly, making it possible to force the victim to release a grip or move their body in a certain way. But there's nothing magical about any of these, and they don't depend on the existence of a magical energy field. Nearly all martial arts call such places on the body as these "pressure points". But many martial arts expand the use of this term to include mythical points along *qi* meridians. So while some pressure points are real, others are hypothetical. An experienced *bullshido* practitioner can certainly give me a series of light blows, even touches, and probably leave me in great pain, possibly even disabled to some degree. He's going to have to actually injure me to do so. What he can't do in reality is what you see in the YouTube videos with his students: Make a series of touches where each, individually, produces no injury; but in conjunction they constitute a disabling attack through manipulation of *qi*. This is the heart of *bullshido*.

Understand and appreciate what martial arts really are and what they can really do. There is science behind why it works; unfortunately it's the prescientific pseudoscientific explanations that are most often repeated. Be aware that martial arts' ancient traditions make them rife with pseudoscience; a fact that continues to be exploited by con artists and clever marketers looking to separate you from your money.

REFERENCES & FURTHER READING

Brett, K. *The Way of the Martial Artist.* Stafford: Donohue Group, Inc., 2008. 75-76.

Gengenbach, M., Hyde, T. *Conservative Management of Sports Injuries.* Sudbury: Jones & Bartlett Learning, 2007. 313-316.

Kelly, M. "Fact or Fiction: Medical Science Examines Dim Mak and its Infamous Death Touch." *Black Belt.* 21 Sep. 2002, Volume 40, Number 9: 79-82.

Maron, B., Gohman, T., Kyle, S., Estes M., Link, M. "Clinical Profile and Spectrum of Commotio Cordis." *Journal of the American Medical Association.* 6 Mar. 2002, Volume 287, Number 9: 1142-1146.

Serino, Danielle. "'The Human Stun Gun' Investigation." *WFLD.* Fox Television Stations, Inc., 4 Mar. 2006. Web. 19 Jan. 2010. <http://www.youtube.com/watch?v=pdrzBL2dHMI>

Thomas, T. *Bruce Lee, Fighting Spirit: A Biography.* San Antonio: Frog Books, 1994. 224-225.

Thompson, P. *Exercise and Sports Cardiology.* New York: McGraw-Hill Professional, 2001. 249-252.

11. THE BELL ISLAND BOOM

Was the cause of this shattering boom in Newfoundland natural, or a test of a superweapon?

In this chapter, we're going to go back in time about a third of a century to 1978, when something strange happened on remote Bell Island in Newfoundland. It was a quiet community, some fishermen and scattered families on small farms and in a smattering of hamlets. It was the last place on Earth you'd expect to be rocked by a sudden, shattering explosion. But it wasn't just an explosion. On what should have been a sleepy Sunday morning, electrical appliances on the island burst apart. Animals fell over dead. Buildings were rent asunder. Though the explosion was the loudest anyone there had ever heard, it seemed to have no precise epicenter. To this day, the cause remains unknown, but hypotheses and conspiracy theories abound. The Bell Island Boom has become, to some people, proof of government tests of secret superweapons.

Newfoundland is a large island off the east coast of Canada, and within Conception Bay at its eastern end is Bell Island. It's a pretty small island, only about 9 kilometers long and 3 kilometers wide. The significant part of its history began in the 1890's when rich iron deposits were discovered and mining began. Tiny Bell Island actually became one of the world's major producers of iron ore, and even saw action in World War II when it was attacked by a German submarine. The mine prospered until the 1960's when pressure from cheaper competition finally forced a closure. Bell Island is low lying with its highest point only a hair above 100 meters, so its deep mines were almost entirely below sea level, requiring constant pumping to keep the water out. So while it was productive, it was always

73

expensive and dangerous. Since the mines closed, they have been filled with water. The island is now laced with tunnels and entrances and passages that lead to underground lakes marking the top of the water table. It's now a site of interest to cave diving adventurers.

With its mine closed and its small economy in shambles, the communities on Bell Island had been largely emptied out for more than a decade when the Bell Island Boom struck on Sunday, April 2, 1978. Without warning, a sudden BANG rocked the island, like a giant electric shock. By some reports, it was heard as far as 100 kilometers away. There was extensive damage to electrical wiring. One property, that of the Bickford family, reported physical damage to their buildings. There were holes in the roof; their television set and fuse boxes actually exploded; and the roof of their chicken coop was destroyed and five chickens inside were killed. Near the shed were two or three holes in the snow that looked like some buried explosives had gone off. Digging in these holes found nothing of interest.

Bell Island

And then the strangest of the stories began trickling in. The Bickford's young grandson reported a hovering ball of light after the blast. One woman on Newfoundland reported a beam of light slanting up into the sky from Bell Island. And several people on the island said they heard a ringing, like a tone, just before the boom struck.

A 2004 documentary film called *The Invisible Machine* postulated that the Bell Island Boom was a test of an electromagnetic pulse weapon. This particular theory was deeply flawed, and depended upon a hypothesized beam weapon being attracted to the iron ore from Bell Island's old mines. These filmmakers were apparently pretty confused to think that natural iron ore is a terrifically powerful magnet, which of course it's not. Although iron is magnetic and can be magnetized, natural iron ore has its molecules jumbled in every direction and rarely happens to have a significant magnetic field, certainly not strong enough to divert or attract a particle beam.

Fueling the fires of conspiracy theories about weapons testing was the mysterious presence of John Warren and Robert Freyman from the Los Alamos National Laboratory (then called the Los Alamos Scientific Laboratory) in New Mexico, who happened to show up right after the boom struck. To many, this was confirmation enough that the Bell Island Boom was indeed a weapons test.

But to understand what the boom really was, we first have to understand the context in which it took place. For about four months before the boom struck, the entire eastern seaboard had been plagued by a series of unexplained booms. Beginning in late 1977 and fading out in mid 1978, some 600 of these so-called "Mystery Booms" left reporters and the public scratching their heads from South Carolina up to Canada. The same conspiracy theories surrounding the Bell Island Boom had been proposed to explain the Mystery Booms. Everything from tests of some kind of superweapon by the United States or the Soviet Union, to an atmospheric doomsday device inspired by Nikola Tesla, or even a massive nuclear device like Convair's proposed Sky Scorcher.

As it turns out, the majority of the Mystery Booms were solved, but with disappointingly mundane explanations that paled beside the much more intriguing conspiracy theories. Many witnesses noted that the Mystery Booms seemed to politely follow a predictable schedule. There was a good reason for this, according to Dr. Gordon MacDonald, who spent seven months compiling an authoritative report on the Mystery Booms for the Mitre Corporation. Of the nearly 600 booms recorded, over 2/3 of them could be positively attributed to sonic booms from aircraft, most notably the Concorde. The Concorde began its flights between New York and Europe only a few days before the first recorded Mystery Boom on December 2, 1977. Because of sonic boom concerns, the Concorde was required to take a wider path out to sea; but on particularly hot days, fuel expansion required the pilots to shortcut over Nova Scotia in order to retain the required fuel reserve. On some days, these booms would propagate up to 100 kilometers by bouncing between two main refracting layers of the stratosphere.

But eliminating the booms that could be traced to supersonic aircraft still left nearly 200 of MacDonald's Mystery Booms unexplained. He attributed them to natural phenomena, but of exactly what nature he did not know. As it turns out, the reason the guys from Los Alamos were there is that they did know. Warren and Freyman had been monitoring imagery from the Vela satellites. This fleet of four satellites kept the globe under constant surveillance, looking for the distinct signatures of nuclear bomb detonations. They also picked up large lightning flashes, and it was in part from the Vela satellites that we learned about lightning superbolts. About five of every ten million bolts of lightning is classified as a superbolt, which is just what it sounds like: An unusually large bolt of lightning, lasting an unusually long time: About a thousandth of a second. Superbolts are almost always in the upper atmosphere, and usually over the oceans. Often these upper atmosphere storms can occur with people on the ground being completely unaware, perceiving nothing other than a clear sunny day. Dr. William Donn, chief atmospheric scientist at Columbia University's Lamont-Doherty Observatory, compiled reports from airline pilots of extremely high-altitude flashes that corresponded with nighttime Mystery Booms. He said the daylight booms may have high-altitude flashes as well, but if they do we wouldn't be able to see them from below because of the daylight.

Although versions of this story on the Internet often claim that Warren and Freyman snooped around and went back to Los Alamos without a word to anyone, the truth is that they spoke quite openly with the media and with other scientists investigating Bell Island. According to Canadian Broadcasting Corporation reporter Rick Cera who spoke with Freyman off-camera, Warren and Freyman had been tracking superbolts over the east coast since December 1977. When they learned of the Bell Island Boom and saw that it correlated with a rare overland superbolt, they were asked to go check it out to see what kind of damage it might have caused. They were surprised the damage to the Bickfords' property was not worse. Warren's specialty was in plasma physics, and Freyman was an expert in RF and its propagation in the upper atmosphere. He's best known as a co-developer of RFID technology. They certainly had a professional interest in observing the aftermath of a superbolt. There's no evidence to support the hypothesis that the Bell Island Boom was the result of a manmade weapon, and all the pieces are in place that support its characterization as a freak lightning superbolt.

Popular tellings of the Bell Island story refer to Warren and Freyman as "military attaches", including *The Invisible Machine* authors. I suppose you could call them this if you're so inclined. Los Alamos is a government facility, but Warren and Freyman were both well-established scientists with publications and legitimate credentials. Whether this qualifies them as "men in black" is, I suppose, a matter of personal language choice. Their visit to Bell Island could reasonably be argued to be consistent with a government weapons plot, but in the same way that my ownership of a toothbrush is consistent with a government plot to keep my breath minty fresh. Imagination is the only evidence supporting either plot.

All the evidence that does exist, however, is consistent with a lightning superbolt. The destruction of electrical appliances was due to the staggering voltage propagating throughout Bell Island's copper power grid. The damage to the chicken shed roof and the electrocuted chickens inside are also consistent with lightning, as were the holes in the snow outside. The angled beam striking the island reported by a witness on Newfoundland could have been a lightning bolt; it's anecdotal and the CBC reporter dismissed it as unreliable. The high-pitched tone preceding the boom reported by witnesses on Bell Island is interesting, but anecdotal; it could have been anything, we can't know, and there's nothing concrete that indicates it was related to the subsequent boom.

The Bickfords' grandson's story of the hovering ball of light is also interesting, and unfortunately anecdotal. Even if we take his story at face value — which stretches the limits of a responsible scientific investigation — we still learn nothing. Hovering balls of light are not consistent with any known phenomenon, certainly not any military weapons, not an electromagnetic pulse, and not lightning either. However, undoubtedly something extraordinary did happen on the Bickfords' property and any report coming from the boy is not surprising. This may constitute an intriguing footnote to the Bell Island Boom, but by no logic can it invalidate the prevailing theory to explain the evidence we have: That a lightning superbolt struck the Bickfords' property.

References & Further Reading

Anonymous. "Tesla Weapon Event at Bell Island, Canada, 1978." *Rumor Mill News Reading Room Archive.* Rumor Mill News Agency, 3 Jun. 2002. Web. 11 Jan. 2010.
<http://www.rumormillnews.com/cgi-bin/archive.cgi?read=20055>

Corliss, W. "Lightning Superbolts Detected by Satellites." *Science Frontiers.* 1 Sep. 1977, Volume 1.

Fleming, A. "The Bell Island Boom." *Suite 101.* Suite101.com Media Inc., 16 Jan. 2010. Web. 17 Sep. 2010.
<http://www.suite101.com/content/the-bell-island-boom-a190154>

Hadley, M. *U-Boats Against Canada: German Submarines in Canadian Waters.* Kingston: McGill-Queen's University Press, 1985. 116, 152.

Stone, J. *Every man should try: Adventures of a public interest activist.* New York: Public Affairs, 1999. 191-198.

Turman, B.N. "Detection of Lightning Superbolts." *Journal of Geophysical Research.* 1 Jan. 1977, 82: 2566-2568.

12. Did Jewish Slaves Build the Pyramids?

It's a popular story, but all the documentary and historical evidence tells us that no Jews were in Egypt at the time of the pyramids.

The stories we hear in Sunday school seem to form the basis for the popular belief that Jewish slaves were forced to build the pyramids in Egypt, but they were saved when they left Egypt in a mass Exodus. That's the story I was raised to believe, and it's what's been repeated innumerable times by Hollywood. In 1956, Charlton Heston as Moses went head to head with Yul Brynner as Pharaoh Ramesses II in *The Ten Commandments*, having been placed into the Nile in a basket as a baby to escape death by Ramesses' edict that all newborn Hebrew sons be killed. More than 40 years later, DreamWorks told the same story in the animated *Prince of Egypt*, and the babies died again.

In 1977, Israeli Prime Minister Menachem Begin visited Egypt's National Museum in Cairo and stated "We built the pyramids." Perhaps to the surprise of a lot of people, this sparked outrage throughout the Egyptian people, proud that *they* had built the pyramids. The belief that Jews built the pyramids may be prominent throughout Christian and Jewish populations, but it's certainly not the way anyone in Egypt remembers things.

Pop culture has a way of blurring pseudohistory and real history, and many people end up never hearing the real history at all; and are left with only the pseudohistory and no reason to doubt it. This is not only unfortunate, it's dangerous. In the words of Primo Levi inscribed front and center inside Berlin's

Holocaust Museum, "It happened, therefore it can happen again." 20th century Jewish history is probably the most important, and hardest learned, lesson that humanity has ever had the misfortune to be dealt. Forgetting or distorting history is always wrong, and is never in anyone's best interest.

I've heard some Christians say the Bible is a literal historical document, thus Jewish slaves built the pyramids (although the Bible doesn't actually mention pyramids, many just assume this); and I've heard some non-religious historians say there's no evidence that there were ever Jews in ancient Egypt. Both can't be true. To find the truth, we need to take a critical look at the archaeological and historical evidence for the history of Jews in Egypt. In order to do this responsibly, we first have to put aside any ideological motivations that would taint our efforts. We're not going to say such research is sacrilegious because it seeks to disprove the Bible or the Torah; we're not going to say such research is a moral imperative because religious accounts are deceptive; and we're not going to pretend that such research is racially motivated against either Jews or Egyptians. We simply want to know what really happened, because true history is vital.

One of the first things you find out is that it's important to get our definitions right. Terms like Jew and Hebrew are thrown around a lot in these histories, and they're not the same thing. A Jew is someone who practices the Jewish religion. A Hebrew is someone who speaks the Hebrew language. An Israelite is a citizen of Israel. A Semite is a member of an ethnic group characterized by any of the Semitic languages including Arabic, Hebrew, Assyrian, and many smaller groups throughout Africa and the Middle East. You can be some or all of these things. An Israelite need not be a Jew, and a Jew need not be a Hebrew. Confusion over the use of these terms complicates research. Hebrews could be well integrated into a non-Jewish society, but modern reporting might refer to them as Jews, which can be significantly misleading.

Now, there are more than just a single question we're trying to answer here. Were the Jews slaves in ancient Egypt? Were the pyramids built by these slaves? Did the Exodus happen as is commonly believed?

The biggest and most obvious evidence — the pyramids themselves — are an easy starting point. Their age is well established. The bulk of the Giza Necropolis, consisting of such famous landmarks as the Great Pyramid of Cheops and the Sphinx, are among Egypt's oldest large pyramids and were completed around 2540 BCE. Most of Egypt's large pyramids were built over a 900 year period from about 2650 BCE to about 1750 BCE.

We also know quite a lot about the labor force that built the pyramids. The best estimates are that 10,000 men spent 30 years building the Great Pyramid. They lived in good housing at the foot of the pyramid, and when they died, they received honored burials in stone tombs near the pyramid in thanks for their contribution. This information is relatively new, as the first of these worker tombs was only discovered in 1990. They ate well and received the best medical care. And, also unlike slaves, they were well paid. The pyramid builders were recruited from poor communities and worked shifts of three months (including farmers who worked during the months when the Nile flooded their farms), distributing the pharaoh's wealth out to where it was needed most. Each day, 21 cattle and 23 sheep were slaughtered to feed the workers, enough for each man to eat meat at least weekly. Virtually every fact about the workers that archaeology has shown us rules out the use of slave labor on the pyramids.

It wasn't until almost 2,000 years after the Great Pyramid received its capstone that the earliest known record shows evidence of Jews in Egypt, and they were neither Hebrews nor Israelites. They were a garrison of soldiers from the Persian Empire, stationed on Elephantine, an island in the Nile, beginning in about 650 BCE. They fought alongside the Pharaoh's soldiers in the Nubian campaign, and later became the principal trade portal between Egypt and Nubia. Their history is known from the Elephantine Papyri discovered in 1903, which are in Aramaic, not Hebrew; and their religious beliefs appear to have been a mixture of Judaism and pagan polytheism. Archival records recovered include proof that they observed Shabbat and Passover, and also records of interfaith marriages. In perhaps the strangest reversal from pop pseudohistory, the papyri include evidence that at least some of the Jewish settlers at Elephantine owned Egyptian slaves.

Other documentation also identifies the Elephantine garrison as the earliest immigration of Jews into Egypt. The Letter of Aristeas, written in Greece in the second century BCE, records that Jews had been sent into Egypt to assist Pharaoh Psammetichus I in his campaign against the Nubians. Psammetichus I ruled Egypt from 664 to 610 BCE, which perfectly matches the archaeological dating of the Elephantine garrison in 650.

If Jews were not in Egypt at the time of the pyramids, what about Israelites or Hebrews? Israel itself did not exist until approximately 1100 BCE when various Semitic tribes joined in Canaan to form a single independent kingdom, at least 600 years after the completion of the last of Egypt's large pyramids. Thus it is not possible for any Israelites to have been in Egypt at the time, either slave or free; as there was not yet any such thing as an Israelite. It was about this same time in history that the earliest evidence of the Hebrew language appeared: The Gezer Calendar, inscribed in limestone, and discovered in 1908. And so the history of Israel is very closely tied to that of Hebrews, and for the past 3,000 years, they've been essentially one culture.

But if neither Jews nor Israelites nor Hebrews were in Egypt until so many centuries after the pyramids were built, how could such a gross historical error become so deeply ingrained in popular knowledge? The story of Jewish slaves building the pyramids originated with Herodotus of Greece in about 450 BCE. He's often called the "Father of History" as he was among the first historians to take the business seriously and thoroughly document his work. Herodotus reported in his Book II of *The Histories* that the pyramids were built in 30 years by 100,000 "workers", which many people have assumed on their own to mean Jewish slaves. Unfortunately, in Herodotus' time, the line between historical fact and historical fiction was a blurry one. The value of the study of history was not so much to preserve history, as it was to furnish material for great tales; and a result, Herodotus was also called the "Father of Lies" and other Greek historians of the period also grouped under the term "liars". Many of Herodotus' writings are considered to be fanciful by modern scholars. Coincidentally, the text of the Book of Exodus was finalized at just about exactly the same time as Herodotus wrote *The Histories*. Obviously, the same information about what had been going on in Egypt 2,000 years before was available to both authors.

Which brings us to the final question: Was there a mass Exodus of Jewish slaves out of Egypt? There is no record of any such thing ever happening, and the simple reason is that there is no time in which it could have happened. No Egyptian record contains a single reference to anything in Exodus; and by the time there were enough Jews living in Egypt to constitute an Exodus, the time of the pyramids was long over. And Pharaoh Ramesses can be let off the hook as well: With apologies to Yul Brynner, no documentary or archaeological evidence links any of the Pharaohs bearing this name with plagues or Jewish slaves or edicts to kill babies. Indeed, the earliest, Ramesses I, wasn't even born until more than a thousand years after the Great Pyramid was completed. His grandson, the great Ramesses II, lived even later.

Some historians have attempted to rationalize the Exodus by drawing parallels to certain cities and trade centers that grew and shrank over the centuries for various reasons. Perhaps one of these economic shifts inspired the story of Exodus. Well, perhaps it did, but the nature of such a migration is, quite obviously, fundamentally different than that depicted in Exodus.

The pseudohistory of ancient Egypt is disrespectful to both Jews and Egyptians. It depicts the Jews as helpless slaves whose only contribution was sweat and broken backs, when in fact the earliest Jewish immigrants were respected allies to the Pharaoh and provided Egypt with a valuable service of both trade and defense. The pseudohistory also takes away from the Egyptians their due credit for construction of humanity's greatest architectural achievement, and portrays them as evil, bloodthirsty slave masters. Pretty much every culture in the world at that period in history included slavery and conflict, and the Egyptians probably weren't any better or worse than most peoples.

Understanding history is essential to understanding ourselves. Although a story like Exodus is profoundly important to so many people throughout the world, the history it describes is false; and the faithful are best advised to seek value in it other than as a mere list of events. Doing so opens the door to a better comprehension of who we are as humans, and it's that shared history that will always unite us — no matter our race, color, or culture. It's just one little more service provided by good science.

References & Further Reading

Awad, M. "Egypt tombs suggest pyramids not built by slaves." *Reuters.* Thomson Reuters, 11 Jan. 2010. Web. 2 Feb. 2010. <http://www.reuters.com/article/idUSTRE6091E720100111>

Comay, J. *The Diaspora Story: The Epic of the Jewish People Among the Nations.* New York: Random House, 1983.

Kraeling, E. *Brooklyn Museum Aramaic Papyri: New Documents of the Fifth Century B.C. From the Jewish Colony At Elephantine.* New York: Arno Press, 1969.

Lindenberger, J., Richards, K. *Ancient Aramaic and Hebrew Letters.* Atlanta: Scholars Press, 1994.

Omer, I. "Investigating the Origin of the Ancient Jewish Community at Elephantine: A Review." *Ancient Sudan-Nubia.* Ibrahim Omer, 1 Jan. 2008. Web. 2 Feb. 2010. <http://www.ancientsudan.org/articles_jewish_elephantine.html>

Porten, B. *Archives from Elephantine: The Life of an Ancient Jewish Military Colony.* Berkeley: University of California Press, 1968.

13. Ball Lightning

We've all heard of it, we all believe it exists - but what does science have to say?

Dive under your desk and take cover: A wicked orb of ball lightning is banging around your room! Or, maybe it's a peaceful, quietly hovering ball of warm light. Or, maybe it's a sparking basketball sized globe chasing you across the plains at night. Or, maybe it's a tiny flaming ball that speeds along and suddenly burns itself out. Whatever it is, it's weird, and it seems to be the first explanation many people will reach for when they hear anything about a  round light source: Ball lightning. What is it? More importantly, is it anything at all? What does science have to say on the matter?

Not much, evidently. And, at the same time, way too much. For as many theories as there are attempting to explain it, there is no agreed-upon description of what they're trying to explain. There are innumerable eyewitness accounts, and almost nothing in common among them. For all the scientists who maintain that it's real, none of them has an accepted theory or any testable evidence. For those who cling to the understanding that ball lightning is indeed an accepted phenomenon, consider these points:

- Ball lightning is not reproducible in the lab. All known forms of electrical discharge are.
- There is no standard description of what ball lightning looks like or how it behaves. Reports of its color, its size, its speed, its sound, the conditions under which it appears, its behavior, its shape, and its duration are all over the map.
- Not a single photograph or video of ball lightning exists that is considered reliable and not otherwise explainable.
- Electromagnetic theory makes no prediction that anything like ball lightning need exist. It does predict all known forms of electrical discharge.

No matter how reliable any one given report might be, it is mired in a sea of other contradictory reports, all describing something very different. This means that either most reports are wrong, or everyone's seeing a different phenomenon. Are some of them actually seeing ball lighting? Maybe, but since we don't know which ones, we don't know what kind of characteristics ball lightning might have; and thus even the anecdotal evidence is too widely at variance to support a single explanation.

In 1997, a reader wrote into *Scientific American's* Ask the Experts column to ask if ball lightning is real. Two experts responded, both giving widely varying descriptions for what it looks like, how it behaves, and where it comes from, but both credulously identifying all such reports as ball lightning. They both had decent sounding hypotheses, though Scientific American referred to them as theories, a status I don't think they've achieved. Both experts, though, displayed what I would consider a red flag. They both speak quite casually using the term "ball lightning" with confidence that it is a real, single phenomenon: Ball lightning has been seen here and here, ball lightning does this or that. In other words, grouping contradictory reports including hoax claims and misidentification of known phenomena all together and explaining them with another unknown, behind which there's no accepted theory.

The first expert, Paul Handel at the University of Missouri at St. Louis, is a long-standing proponent of the hypothesis that ball lightning is a manifestation of a maser caused by regular lightning striking within a standing wave of UHF or microwave radiation. In 1975 he developed what he calls "Maser-Soliton Theory" to describe this. From his description in the *Scientific American* column:

> ...*The maser is generated by a population inversion induced in the rotational energy levels of the water molecules by the short field pulse associated with streak lightning. The large volume of air that is affected by the strike makes it difficult for photons to escape before they cause 'microwave amplification by stimulated emission of radiation'... Unless the volume of air is very large or else is enclosed in a conducting cavity, ...collisions between the molecules will consume all the energy of the population inversion. If the volume is large, the maser can generate a localized electrical field or soliton that gives rise to the observed ball lightning. Such a discharge has not yet been created in the laboratory, however.*

If you find yourself asking "What the heck is he talking about?" you're not alone. As very few scientists outside of some Russian colleagues of Handel's have written about his "Maser-Soliton Theory", it's fair to say that it appears he has yet to convince any significant number of scientists of its validity. The requirement that there happen to be a standing wave of electromagnetic radiation (of unknown origin) when the lightning decides to strike is one reason.

Handel is not the only one pointing at microwaves, though. The Internet is full of instructions for creating ball lightning in your microwave oven, none of which I recommend that you attempt. Placing carbon veil or fine steel wool in a microwave oven will create a glowing plasma that will damage the roof of your oven unless contained within Pyrex. Burning a candle flame with carbon pencil rods or carbon charred toothpicks will produce a similar effect. But referring to these kitchen experiments as "ball lightning" is a bit of a strain. First, the plasma created is not shaped like a ball. Second, being extremely hot (dangerously hot), it rises upward, which is a behavior rarely seen in ball lightning reports. Third, it requires a high-powered microwave oven doing its thing, which probably explains why Paul Handel's "Maser-Soliton Theory" has not produced observable effects in nature.

The second expert who offered his thoughts in *Scientific American* was John Lowke at Australia's Institute of Industrial Technologies. He proposed the mechanism to be the rapid discharge of electrical energy from a lightning bolt that has struck the ground. As the electrical charge disperses through the ground, it creates a plasma similar to the more familiar corona discharge of St. Elmo's Fire. He proposed that the movement of the ball would be determined by the speed at which the charge moves through the ground, which could explain why some reports state the ball lightning moved against the direction of the wind. But once again, nobody has ever been able to produce this effect artificially, and Lowke acknowledged that "There is no generally accepted theory of ball lightning."

The corona discharge hypothesis is the most interesting, as St. Elmo's Fire is a well understood and well established phenomenon with a sound underlying theory. It's the same thing that makes a fluorescent light glow. When there's a big difference between the electrical charge in the ground and the atmosphere, electrons flow from one to the other. They do this most efficiently out the tips of sharp conductive points; masts on a ship being the most familiar example. Given a strong enough field off this tip, the air is turned into a plasma that fluoresces. St. Elmo's Fire is blue or purple in air. If our atmosphere was neon, it would be reddish orange, and so on for all the other colors that fluorescent tubes come in.

But St. Elmo's Fire has an obvious power source: The powerful flow of electrons coming from the conductive point. Ball lightning, while descriptions of its color are often similar to that of St. Elmo's Fire, has no apparent power source. This might make the microwave hypothesis more attractive, but we have no theory that would explain the concentration of the effect in a sphere, and no theory to explain why there might happen to be a standing microwave. Making light requires energy. Any valid theory of ball lightning has to include the power source for all that light.

No discussion of ball lightning, or any other electrical phenomenon for that matter, is complete with the obligatory mention of the patron saint of eccentric electrical theorists, Nikola Tesla. The popular rumor you always hear is that Tesla was able to produce ball lightning at will in his lab. Regarding what he called "electric fireballs," Tesla reported in 1904 in the journal *Electrical World and Engineer* "I have succeeded in determining the mode of their formation and producing them artificially." Sadly for the world of science, Tesla's own claims on this matter were never evidenced and have never had any reliable corroboration. There's one oft-repeated quote attributed to Tesla, which seems to be a proposed explanation for fireballs he observed and hoped to recreate:

> *...It became apparent that the fireballs resulted from the interaction of two frequencies.... This condition acts as a trigger which may cause the total energy of the powerful longer wave to be discharged in a infinitesimally small interval of time... and is released into surrounding space with inconceivable violence. It is but a step, from the learning how a high frequency current can explosively discharge a lower frequency current, to using the principle to design a system in which these explosions can be produced by intent.*

Separately, in his Colorado Springs Notes, Tesla attributed ball lightning to resistively heated particles in the air. Just as a light bulb's filament produces heat and light from electrical resistance, so might a carbon particle in the air if exposed to high current. It's a fine speculation, but such a fireball would rise and flame out rapidly (like the plasma created in a microwave), it would not hold a ball shape and hover; even if it did, it would require an extraordinary power source and the presence of carbon particles floating about. That's inconsistent with most ball lightning reports, as are explosions of "inconceivable violence". So really none of what Tesla reported bears much similarity to the ball lightning reports that we commonly hear.

So then, in summary, what about this popular trend of suggesting ball lightning as an explanation for a strange report of a hovering ball of light? It's a little hard to justify. As ball lightning has no established properties, it cannot be argued to be a probable match for any given report. It is fair to say that it's likely that one or more unknown phenomena exist that have triggered eyewitness accounts of hovering balls of light, but there's insufficient theory to support assigning these accounts a positive identification of ball lightning. Indeed, as ball lightning can only honestly be described as an unknown, it would be illogical to use it as an explanation for any report.

References & Further Reading

Barry, J. *Ball Lightning and Bead Lighting: extreme forms of atmospheric electricity.* New York: Plenum Press, 1980.

Handwerk, B. "Ball Lightning: A Shocking Scientific Mystery." *National Geographic News.* National Geographic Society, 31 May 2006. Web. 25 May. 2011.
<http://news.nationalgeographic.com/news/2006/05/060531-ball-lightning.html>

Stenhoff, Mark. *Ball Lightning: An Unsolved Problem in Atmospheric Physics.* New York: Kluwer Academic/Plenum Publishers, 1999.

Tesla, N. *Colorado Springs Notes.* Beograd: Nolit, 1978. 333.

Trefil, J. *Ball Lightning, UFOs, and Other Strange Things in the Sky.* New York: Scribner's, 1987.

Uman, M., Handel, P. "Ask the Experts." *Scientific American.* Scientific American, 18 Jul. 1997. Web. 5 Mar. 2010.
<http://www.scientificamerican.com/article.cfm?id=periodically-i-hear-stori>

14. The Faces of Bélmez

Faces appearing on the floor of a house in Spain were easily proven to have been faked – why does their story carry so much ongoing influence?

You can hardly find the village of Bélmez de la Moraleda for all the olive groves it's hidden among. Driving through the foothills in this part of southern Spain, you see olives, olives, and more olives. But turn up the slope and you'll discover its white stucco buildings drenched in sunlight, its 2,000 residents hard at work keeping it spotless and neatly trimmed. But not everything in this pastoral hamlet is quite so cheery: For more than 30 years, one house in town enclosed a frightening portent. The image of a face once appeared on its concrete floor. Its owners tried to remove it, but it persisted. And soon it was joined by others. And others. Soon, ghastly, ghostly countenances were everywhere on the floor, sometimes even overlapping. Small faces, big faces, some benevolent, some shrieking in horror. They became known as the Faces of Bélmez.

Now I'm going to spare you ten minutes of discussion and investigation into a case as cut and dried as this one. My point today is not to "debunk" the Faces of Bélmez, but rather to look at its significance in a broader context. So I'm just going to give you a quick and dirty reveal: The faces were shown to have been painted on the concrete floor, the first with paint and later with acid, and the woman living in the house found to be perpetrating a hoax on the public for financial gain. In 1971, María Gómez Cámara announced the appearance of the first face on the kitchen floor, and psychic believers called it a case of "thoughtography", claiming that Cámara's thoughts had a telekinetic effect and projected images from her mind onto the floor. When she and her family began charging admission for tourists to see the faces, the mayor had a sample removed for testing. The hoax was easily revealed, and the city banned any further tourist business from being conducted at the residence. However that did not stop them, and the faces continued to appear for more than 30 years until Cámara's death in 2004.

When I give a public lecture, one of the first questions I'm often asked is why do people believe this stuff? And though it would be nice to have "a" reason, the truth is that there are probably as many different reasons as there are people. But there are some common themes. When we look at the Faces of Bélmez, you and I see something that's an obvious hoax, that it seems you'd have to be unrealistically ignorant to believe in. Ignorance probably plays a role for some people, but for most of the pilgrims who traveled to see the faces, cultural context was probably a bigger factor. We're not surprised to learn about Native American spiritualism (for example), and neither should we be surprised to learn of European traditions. The Andalusian region of Spain places deep emphasis on the Virgin Mary, and the lower classes in particular bear the influence of eastern European gypsies, called *Romani gitanos*. Even the *gitanos*' style of dress shows obvious gypsy cues. Among gypsy traditions are a deeply rooted belief in psychic powers, faith healing, and communication with the dead. Thus we have a population molded by a blending of Catholic miracles with psychic powers; and to many of these people, the Faces of Bélmez are a practically inevitable confirmation of *Santa María*.

In a larger context, the prospect of finding meaning within everyday things is compelling to most people. We want the things we do and see to have a deeper dimension that suggests the existence of a power greater than what we can observe. We want to be able to have that power too. When we see that Cámara can turn her thoughts into tangible reality, we strongly want it to be true, in part because of how attractive is the prospect of being able to do that ourselves; but also because it's spiritually comforting to know that such powers are out there watching over us.

Perhaps a simpler question is why would someone perpetrate such a hoax on a trusting public? In this case, the city government found that it was as simple as financial gain. But in my experience, people like Cámara are rarely simply hustlers. If she was like most career psychics, she probably believed her powers were real. And even as she and her family took paintbrushes in hand and deliberately faked the faces on the floor, cognitive dissonance allowed her to still believe that what she was doing was real. She could have honestly believed that she'd been divinely inspired through psychic abilities to paint the faces, and it could be as simple as that. She could well have believed that the psychic advice she dispensed throughout the village was just as inspired. There are so many possibilities, both including and excluding conscious fraud, that it's premature to make any determinations about her character or motivations.

One of the Faces of Bélmez

And so, even giving the benefit of the doubt and assuming only the best of motivations for Cámara and others like her, are any ethical implications nullified? I argue that they're not. Until psychic abilities can pass any kind of controlled test and be shown to exist, it's safe to say that Cámara's psychic advice is no better than what you might get from a cat, or the flip of a coin. Knowingly or not, Cámara violated any kind of ethical code you might choose to apply to what she did. At a minimum, she allowed the belief to persist that the faces appeared spontaneously. She could easily have painted them on canvas or anywhere else and offered the story that they were the result of psychic abilities, but she didn't. She stuck to what she knew was a lie, she derived profit from it, and she knew that people took her advice in part because of the bolstering provided by the lie.

The same can be said of anyone who dispenses a product or service who has good reason to doubt the value of such a product. Every psychic, homeopath, and acupuncturist is well aware that controversy exists regarding the validity of their product. Even though they may have thoroughly convinced themselves that their products work, they know that not everyone agrees, and they know that the overwhelming body of empirical evidence is against them. Every ethical practitioner should refuse to accept another penny until they can establish that their service is indeed of real value to the customer. When doctors learn that a certain drug is found to be worthless, they should stop prescribing it, regardless of their personal feelings about it. Psychics should have no less of an ethical obligation.

Perhaps more important than the ethical implications of tricks like this are its intellectual ramifications. Believing tourists who come to Bélmez to see the faces are transformed from innocent believers who take the story on faith, into experienced witnesses who have evaluated the evidence firsthand and no longer need the faith. What was merely a superstitious belief is now authoritative knowledge of how the universe works. And, it's wrong. When you validate someone's superstitious belief, you teach them that other superstitions are likely true as well. Someone who witnesses the proof of a magical claim firsthand is much more likely (perhaps even certain) to fully embrace any other claims coming from the same source or similar sources.

Consider the possibility of a child from one of these villages getting an infection or diarrhea. Left untreated, or ineffectually treated, these can be fatal; when a simple trip to any doctor could easily cure them. The Faces of Bélmez teaches villagers to put their faith instead into a traditional treatment left over from the days when these conditions were virtual death sentences.

But of course, it's relatively rare that someone's superstitious beliefs actually put their life in danger. What's much more common is the reliance on psychic information to guide decisions in business or personal relationships. By all accounts, María Gómez Cámara was among the most influential people in Bélmez. People brought her questions every day, seeking what they believed was guidance divinely inspired by *Santa María*. The city council's determination that her painted faces were fraudulent apparently had no effect on the public's perception of her abilities. From what we know, it appears that nearly every day in Bélmez during Cámara's lifetime, some significant decision was made based on information of totally unknown value. I found no record indicating whether the advice she dispensed turned out to be right or wrong, so there's no foundation to assert that her advice was not divinely inspired or was not of faultless quality. We have only the one data point: The faces she and her family painted on the floor, and then fraudulently presented as evidence of her divinely inspired psychic abilities. Is it possible that her abilities were otherwise truly psychic? Certainly. Is it likely? Judge for yourself.

I argue that the promotion of bogus phenomena like the Faces of Bélmez instills a harmful lesson into people that damages their ability to be successful in many facets of life. A more valuable lesson would be the critical thinking skills that allow them to properly analyze such claims, and discriminate true information from false. I often hear the question put forward, is it important to "debunk" stories like this, even if people draw comfort and hope from them? Based on the point I just made, yes, without question; but mere "debunking" is of no practical value if left there. What you need to do is teach people the critical thinking skills that will help them avoid such pitfalls in the future. That's what's important.

So, the Faces of Bélmez: A harmless little folk tale? Perhaps, but only if taken out of context and considered by itself. But within the context of the lives of the believers who come to see it, it is unambiguously a harmful influence, contributing to the erosion of their ability to make sound decisions governing their lives.

REFERENCES & FURTHER READING

Baker, P. "History Mystery: The Bélmez Faces." *Helium.* Helium, Inc., 1 Mar. 2010. Web. 27 Nov. 2010.
<http://www.helium.com/items/1899203-history-mystery-the-belmez-faces>

Forte, R. "Faces of the Dead Mysteriously Appear on Woman's Cement Floor." *Weekly World News.* 10 Oct. 1995, Volume 17, Number 2: 40-41.

Nickell, J. *Looking for a Miracle; Weeping icons, relics, stigmata, visions & healing cures.* Amherst: Prometheus Books, 1993.

Schweimler, D. "An unexplained mystery." *News.bbc.co.uk.* BBC Online Network, 19 Oct. 1999. Web. 7 Oct. 2010.
<http://news.bbc.co.uk/2/hi/programmes/from_our_own_correspondent/476045.stm>

Tort, C. "Bélmez Faces Turned Out to Be Suspiciously 'Picturelike' Images." *Skeptical Inquirer.* 1 Mar. 1995, Volume 19, Number 2: 4.

Wynn, C., Wiggins, A. *Quantum Leaps in the Wrong Direction: Where real science ends...and Pseudoscience Begins.* Washington, DC: Joseph Henry Press, 2001.

15. The Denver Airport Conspiracy

Is the Denver International Airport really a headquarters for the New World Order?

On the surface, Denver International Airport seems like any other modern airport. It's new, it's clean, it's big, and it's modern. But some investigators have found more to it than meets the eye. Much more. Claims abound that Denver International was designed and built by the Illuminati as the headquarters for the global genocide that will trigger the New World Order.

An Internet search for "Denver airport conspiracy" reveals that there is a lot of talk about this, and that the specific claims and observations are numerous. Here's an overview of the basics. According to the conspiracy theorists, Denver already had a fine airport, Stapleton International. But despite widespread protests, Denver International was built and opened in 1995, with fewer runways, thus *reducing* Denver's capacity. Its construction began with five mysterious buildings that were completed and then buried intact, with the cover story that they were "built wrong". Up to 8 levels of underground facilities are said to exist, and workers who go there refuse to answer questions about what they do. The entire airport is surrounded by a barbed wire guard fence, with the barbed wire angled inward, to keep people *in*, like a giant prison, not *out* like at other airports. And if viewed from the air, the runways are revealed to be laid out in the shape of a Nazi swastika. Questions about what the government might be doing in this underground base may have been answered in 2007, when fourteen commercial aircraft reported spontaneously shattered windshields as the presumed result of electromagnetic pulses.

Indoors, the airport gets even stranger. The Illuminati appear to have detailed their plans for global genocide and a New World Order in two large murals. The first depicts a huge Nazi soldier with dead women and children scattered around him, and the second shows Third World populations dying, a few elite species protected from the apocalypse in sealed containers, and the Mayan symbol for 2012 presiding over all. In the floor near the murals is written "Au Ag", the abbreviation for the deadly toxin Australia Antigen, evidently the Illuminati's weapon of choice to accomplish the genocide. On other places, strange words in an unknown (possibly alien) language are written on the floors ("DZIT DIT GAII" and others). Most telling of all is the granite monument that the airport claims is a time capsule. It's emblazoned with the symbol of the Freemasons, well known among conspiracy theorists to be a major arm of the Illuminati, and engraved with the words "New World Airport Commission". And finally, the Queen of England, another alleged Illuminatus, has been secretly and anonymously buying up the property surrounding the airport.

Some of the "alien language"

There's a common red flag shared by this particular conspiracy and many others, and that's the presumption that the conspirators chose to publicly announce their evil plans by putting all of this out there for everyone to see. That would be like Nixon, before Watergate, ordering a public mural to be placed in the hotel lobby showing GOP spies breaking into a room.

Or Oliver North announcing his intentions by placing a sculpture in the National Mall showing himself handing a shoulder-fired missile launcher to an Iranian with one hand, and giving the proceeds to a Nicaraguan with the other. If these analogies sound absurd, they are; but they do accurately portray the conspiracy theorists' belief of why Denver International has certain features: it was to publicly announce plans for a future crime against humanity. If you don't think Nixon or North were fool enough to make such announcements, you might want to reconsider whether that's actually what was done at Denver International.

Was the old Stapleton International Airport truly adequate, and did it really have more capacity? Not even close. It was 65 years old, was a major noise nuisance being in the middle of the city, and had only three 10,000-foot runways, the minimum needed for large jets at such a high elevation, and barely adequate for fully loaded international jumbos. Denver International has six runways all over 12,000 feet, and one at 16,000 feet, that can accommodate any jet in the world. Its location clear of Denver alleviates the noise concerns.

The underground constructions, as anyone who has traveled through Denver knows, are for the underground train system that connects all the terminals, including additional tunneling built to accommodate future expansion. Other underground systems were built for Denver's state of the art automated baggage handling system. Unfortunately the system never worked well, and by 2005 it was retired completely, and now the underground tunnels are used for conventional baggage handling. Hundreds of workers go in and out of there every day, and none of them have ever reported seeing anything unusual. No reptoids, no aliens, no Illuminati. Yet. I was not able to find any well-reported cases of a worker being interviewed and refusing to discuss his job, so as far as I can tell, this was simply made up.

What of the large barbed wire prison area? Like those at all commercial airports in developed nations, Denver International's runways are fenced off with barbed wire. This is required by law for obvious safety reasons. Stories that the barbed wire is angled inward are simply untrue: People who have actually been there report that the barbed wire sticks straight up and is not angled either way. There appears to be no significant difference between the fencing at Denver International and that at any other commercial airport.

As for those broken windshields, it really did happen. It took the newspapers a little while to get the cause from the National Transportation Safety Board, and that period of uncertainty is when the rumors about electromagnetic weapons germinated. The NTSB confirmed what the mechanics who replaced the windshields were able to tell at a glance: The breaks were caused by flying debris during high winds.

Whether the runways are laid out like a swastika seems to be a matter of opinion. It's easy to see on Google Earth, and to me it certainly does give the general impression of a swastika. If you select certain taxiways and runways, you can overlay a near-perfect swastika, but to do so you have to ignore very significant runways including the biggest one. I'd say the eastern half does look like half a swastika, but the western half is too much of a stretch. Even if it was perfect, I don't see what that would prove. Runways are laid out with prevailing winds and have to be approved by safety agencies and pilot groups. If a swastika is what works, a swastika is what they'll use.

The swastika theory was bolstered by the Nazi-like figure in those artistic murals depicting, supposedly, the Illuminati's plans to reduce the Earth's population. In fact those two creepy murals were each half of a diptych, a two-part mural, each depicting a hopeful message of man's journey from brutality to peace. They were made by Chicano artist Leo Tanguma, one of several artists commissioned to paint similar murals throughout Denver International. The *Children of the World Dream of Peace* shows a menacing Nazi-style soldier wreaking havoc and fear, and includes a poem written by an actual child who died at Auschwitz. In the second half of this diptych, the soldier is dead, and the children of all nations come together over his corpse and beat the world's swords into plowshares, inspired by the Bible verses Isaiah 2:4 and Micah 4:3.

Tanguma's other mural, *In Peace and Harmony with Nature*, shows the Earth suffering from exploitation, and some species extinct and now found only behind museum glass, mistaken by the conspiracy theorists for elite species being protected from the Apocalypse. The alleged Mayan reference to 2012 is simply a small piece of carved stone with Mayan-style decorations held by one figure in the mural and representing the decline of indigenous populations. (Tanguma himself is a Mayan.) The second half of this diptych shows children from all nations gathering at the flowering tree of peace on a rejuvenated Earth.

But the conspiracy theorists have been told all this; they simply choose to regard it as untrue; in part because of that warning etched into the floor near Tanguma's murals that Australia Antigen will be used to kill us all. Colorado is well known for its history of gold and silver mines, and so a mine car full of Au and Ag is a most appropriate symbol for the state. Is it also an appropriate symbol for a planned genocide by the Illuminati? Well, a mine car doesn't seem to have any particular significance to genocide; if the mine car was put there to confuse us and throw us off the trail, and the idea was for the clue to be obfuscated, why give the clue at all? Australia Antigen is indeed sometimes abbreviated as Au Ag, but its true common abbreviation is HBsAg, for Hepatitis B antigen. By itself it would not be much of a biological weapon. If you exposed everyone to Australia Antigen, you'd essentially just inoculate them against Hepatitis B. An antigen is not the disease germ itself, it is simply any substance that provokes an immune response. Australia Antigen is merely the protein found on the surface of the Hepatitis B virus. By itself, it's harmless. A vaccine has existed for some time against Hepatitis B, and many people have received it. Even if you haven't, Hepatitis B is treatable and is rarely fatal; and so, again, a pretty poor choice for a mechanism of global genocide.

The other words in the floor (the "alien language") are Navajo names for sacred mountains in Colorado, once again, a significant part of the state's history.

And then there's that granite 100-year time capsule bearing the symbol of the Freemasons on its capstone. Could there be any explanation other than Denver International being the headquarters of the Illuminati? Another explanation might be that the local Masonic lodge constructed and placed the time capsule. The Masons are a civic organization that often performs such services, and this has been a tradition at public buildings in the United States for a long time. Both the White House and the US Capitol have cornerstones placed and marked by local Masonic lodges. Although you might choose to interpret this to mean that Denver International is some kind of government headquarters, you could also observe that these are just three examples of hundreds, if not thousands, of public buildings in the country that followed this tradition.

The "New World Airport Commission" mentioned on the capstone did exist, contrary to conspiracy claims that there is no record of any such organization. The commission was a group of Denver business and civic leaders who sponsored and organized some events at the airport's opening, ushering in Denver as a new "world-class" city. You can say this is false, but all you've got to back that up is your own imagination.

Perhaps the most entertaining of the conspiracy claims about Denver International Airport concerns the brass plaque at the end of an artistic curved pedestal above the time capsule. The plaque repeats the Mason symbol and has the rest of the text from the time capsule written in Braille. However, many conspiracy web sites refer to this paragraph of Braille as a "keypad". I even found a lengthy discussion on AboveTopSecret.com, the notorious conspiracy theory web site, speculating about what the code might be. In all seriousness, some seem to believe that a code can be typed on this solid plaque which will either open the time capsule, permit admittance to the secret underground base, or perhaps even trigger the global genocide.

Queen Elizabeth II is indeed a pretty substantial property investor, although it's really the Royal Crown in general and not her personally. Her own fortune is a few hundred million dollars, but the majority of the Crown's wealth is held in trust for the nation. This portion is somewhere north of $17 billion and includes property investments all around the world. Although neither I nor a friend in the real estate department at a Denver law firm could find any specific record of the Crown purchasing land near Denver Airport, real estate near a new international airport is usually a sound investment for anyone, and such a purchase would be consistent with the Crown's other investments.

Of course, if we don't uncritically accept all this, we're "sheeple" who have had the wool pulled over our eyes by the Illuminati overlords, and I'm merely a shill paid to spread misinformation (but you knew that about me already). The truth is that the Apocalypse is coming, and the New World Order has chosen to publicly announce their plans in an airport; but only the specially gifted "Patriots" are enlightened enough to see it.

References & Further Reading

Bingham, J., Singh, A. "Queen 12th in Forbes list of richest royals." *Daily Telegraph*. 21 Aug. 2008, 2598399: 8.

DIA. "Art at DIA." *Denver International Airport*. City & County of Denver Department of Aviation, 5 Feb. 2003. Web. 23 Feb. 2010. <http://www.flydenver.com/guide/art/>

Editors. "DIA Cracked Plane Windshields Mystery Solved." *TheDenverChannel.com*. KMGH-TV, 28 Feb. 2007. Web. 23 Feb. 2010.
<http://www.thedenverchannel.com/news/11133070/detail.html>

Glass, R. *Software Runaways*. New York: Prentice Hall, 1998.

Google. "Denver International Airport." *Google Maps*. Google.com, 9 Jun. 2007. Web. 23 Feb. 2010. <http://is.gd/aKwB4>

Li, A., Williamson, R., et al. "New Denver Airport: Impact of the Delayed Baggage System -- GAO/RCED-95-35BR." *Research and Innovative Technology Administration.* US Department of Transportation, 14 Oct. 1994. Web. 23 Feb. 2010. <http://ntl.bts.gov/DOCS/rc9535br.html>

16. Zeitgeist: The Movie, Myths, and Motivations

The Internet movie Zeitgeist uses flagrant dishonesty to make an ideological point that could have easily been made ethically.

In this chapter, we're going to point the skeptical eye at one of the most popular Internet phenomena from the last couple of years: *Zeitgeist*, a freely downloadable documentary movie. It purports to critically examine Christianity, the cause of 9/11, and the world economy. Instead, it paints them all with a single wide stroke of the conspiracy paintbrush. "Zeitgeist" is a German word meaning the spirit of the times, thus *Zeitgeist* the movie purports to pull aside the curtain and reveal the true nature of the world in which we live. The problem with the film, as has been roundly pointed out by academics worldwide, is that many of the conspiratorial claims and historical references are outright fictional inventions. *Zeitgeist* does have a message that's not necessarily invalid, but it's lost underneath the unequivocal dishonesty.

For a long time, people have been asking me to turn Skeptoid's skeptical eye on *Zeitgeist*. I've resisted, mainly because it's so poorly researched that I didn't feel it deserved any response from legitimate science journalism. But people have kept asking. And, obviously, a lot of viewers have been swayed by it. I've even had people who innocently bought into it write me and quote *Zeitgeist* as an authority, suggesting I do some episode promoting one of its claims. *Zeitgeist*, and the 9/11 conspiracy movie *Loose Change*, are largely what motivated me to produce *Here Be Dragons*, my free 40-minute video giving a general introduction to applied critical thinking, which I felt was a more appropriate response than publicly acknowledging either film. But I spent some time learning more about *Zeitgeist*, its sequels and related events, and its creator, and concluded that the mainstream criticism of the film doesn't tell the whole story, and its worldwide impact does make it deserving of a more critical examination.

Understanding *Zeitgeist* means understanding its creator, Peter Joseph Merola (credited as Peter Joseph), a young musician, artist, and freelance film editor living in New York City, at last account. I've found no reference to any educational or professional experience pertaining to any of the subjects covered in the movie. He moved to New York in order to attend art school. That appears to be the extent of his qualifications to teach history and political science, but of course it doesn't make him wrong. It may, however, explain why many of his factual claims contradict what anyone can learn from any textbook on religious history or political science.

Joseph made a second film, *Zeitgeist: Addendum* which offers much better insight into the man and his motivations for creating *Zeitgeist*. He's basically a postmodern utopian, who spends most of his effort speaking out against money-based economics. He advocates the rejection of government, profit, banking, and civil infrastructure: basically, the "establishment". Once you understand where he's coming from, it makes it a lot easier to understand why he made *Zeitgeist* and tried so hard to point out the corruption and evils of the establishment. The problem is that he simply made up a bunch of crap to drive his point, and that's where he crossed the line between philosophical advocacy and unethical propaganda.

Much of what makes *Zeitgeist* popular is that the sustainable utopia he describes is very compelling. It's probably not very realistic, but it's alluring at an organic level. Mistrust of the establishment has been a popular theme ever since a caveman first raised a club, so the two combine to make the message of *Zeitgeist* appealing, at some level, to nearly everyone. For example, in his sequel, Joseph profiles futurist Jacque Fresco who envisions what he calls a "resource-based economy", a world without money where the Earth's natural resources are freely available to all and responsibly managed through public virtue and high technology. This is a fine idea, and while its practicality and workability can certainly be debated, it's perfectly valid as a philosophy. And so, it was from this utopian perspective that the young idealist Peter Joseph Merola set out to first convince us that our current system is fundamentally broken.

He began in the first of *Zeitgeist's* three chapters with an assault on Christianity. The film draws many parallels between the Nativity story and pagan sun worship and astrology, suggesting that their origins are all the same. This is followed by an impressive set of similarities between the life of Jesus and the life of Horus, the Egyptian god — similarities far too extensive to be simple coincidences. And then, taking key points from the life of Jesus (the virgin birth, December 25th, a resurrection after three days, and so on), we find that the same elements are found in the stories of many other gods from diverse cultures, namely the Phrygian Attis, the Indian Krishna, the Greek Dionysus, and the Persian Mithra. Joseph's presentation is compelling, and constitutes a convincing argument that Christianity is just one of many branches of mythology stemming from the same ancient stories going all the way back to prehistoric sun worship.

Where this compelling presentation breaks down is, well, almost everywhere. The majority of Joseph's assertions are flagrantly wrong, as if he had begun with a conclusion, and worked backwards making up facts that would get him there. He gave no sources, but it turns out that most of these same claims about other gods having the same details as the Jesus stories come from a 1999 book called *The Christ Conspiracy: The Greatest Story Ever Sold*. Christian scholars in particular have been highly critical of Joseph's unresearched and wrong assertions, which is understandable given that they are probably the best authorities on religious histories.

Part II of the movie depicts the 9/11 attacks as having been perpetrated by the American government, essentially repeating the same basic charges found throughout the 9/11 "truth" community. These charges fall into two basic categories: innuendo and misinformation. Innuendo like the Bushes knew the bin Ladens, the alleged hijackers have since been found to be alive and well, the inexperienced pilot couldn't have hit the building; and misinformation like straw man arguments mischaracterizing what we all watched that day. These, and many other tactics claimed by the "truthers" to be evidence that the attack was an inside job, have been thoroughly addressed elsewhere and I'm not going to go into them here. In short, searching for alternative possible motivations, and finding and making extraneous connections between various people and events, does not prove or serve as evidence of anything. Raising the specter of doubts or alternate possibilities is very effective in distracting people away from the facts, as we saw so dramatically in O. J. Simpson's murder acquittal, and as we see throughout the 9/11 "truth" movement.

According to a New York Times interview with Peter Joseph Merola in which he was asked about the 9/11 conspiracy claims made in *Zeitgeist*, he says he has since "moved away from" these beliefs. While it's great that he was willing to come out publicly and say that he's abandoned one line of irrational thinking, to me it says more that he leaves it in the movie anyway *(Zeitgeist* has gone through a number of revisions, and he's had ample opportunity to edit out sections he no longer believes). This is only speculation on my part, of course, but I'd guess he leaves it in because it so dramatically illustrates the evils of the establishment, which is a pillar of his philosophy. If true, it would show that the content of Joseph's films are driven more by ideology than by fact.

That this is Joseph's ideology is most impactfully illustrated in part III of *Zeitgeist*. This asserts the existence of what Joseph believes is a worldwide conspiracy of international bankers, who are directly responsible for causing all wars in the past century as a way to earn profits. From his student art studio, Joseph purports to have uncovered plans, known only to a select few of these hypothesized bankers, to combine the currencies of Canada, the United States, and Mexico into a single denomination called the amero, as a next step toward an eventual one world government. In fact, the amero was proposed in a couple of books: in 1999 by Canadian economist Herb Grubel in *The Case for the Amero*, and in 2001 by political science professor Robert Pastor in *Toward a North American Community*. The number of economists *not* proposing an amero is much larger. This chapter of *Zeitgeist* goes into great detail, most annoyingly in the way it quote-mines everyone from Thomas Jefferson to Carl Sagan (from letters both real and counterfeit) to suggest that leaders in government and science have always known about this. People knowledgeable in this subject have gone through Zeitgeist point-by-point and refuted each and every one of its dishonest claims, none more effectively than Edward Winston on his Conspiracy Science web site, which I highly recommend if you want to discuss any of the nitty gritty details in any section of *Zeitgeist*.

I can empathize with Peter Joseph Merola on one level. When I first started the Skeptoid podcast, I didn't really yet know what it was going to be about or where it was going to lead. I didn't keep references either. Having done it a few years, I now have my focus dialed in much better. I can see the same evolution from the conspiracy theories in the original *Zeitgeist* film to the utopian and philosophical topics Joseph now talks about. He described *Zeitgeist's* inception as a personal project and a "public awareness expression", a context in which it was unnecessary to keep references or even to be historically accurate. I suspect that if he'd known where he was going to be today, he wouldn't have made *Zeitgeist*, and would have instead gone straight to the sequel which almost completely omits the conspiracy theories and untrue history.

If he had, the *Zeitgeist* franchise would probably not be nearly so successful. Nothing commands attention and feeds our native desire for power like a good conspiracy theory. If you know about the conspiracy, you're in on the secret information, and you are more powerful than the conspirators. For better or for worse, we all have a deep craving to have the upper hand. This is perhaps the main reason for the unending popularity of *Zeitgeist, Loose Change,* Alex Jones, Richard Hoagland, and other conspiracy theory machines. It also explains the passion shown by those who defend them: All that matters is "being the one who knows more than you," and the facts are a distant second.

References & Further Reading

Callahan, T. "The Greatest Story Ever Garbled." *Skeptic.* The Skeptics Society, 25 Feb. 2009. Web. 2 Mar. 2010. <http://www.skeptic.com/eskeptic/09-02-25>

Dunbar, D., Reagan, B. *Debunking 9/11 Myths: Why Conspiracy Theories Can't Stand Up to the Facts.* New York: Hearst Books, 2006.

Feuer, A. "They've Seen the Future and Dislike the Present." *New York Times.* 16 Mar. 2009, N/A: A24.

Lippard, J. "Zeitgeist: The Movie." *The Lippard Blog.* Jim Lippard, 11 Jun. 2008. Web. 2 Mar. 2010. <http://lippard.blogspot.com/2008/06/zeitgeist-movie.html>

Meigs, J. "Debunking the 9/11 Myths: Special Report." *Popular Mechanics, March 2005 Issue.* 1 Mar. 2005, Year 103, Number 3.

Pastor, Robert A. *Toward a North American Community: Lessons from the Old World for the New.* Washington: Institute for International Economics, 2001. 111-115.

Siegel, Jon. "Income Tax: Voluntary or Mandatory?" *Jon Siegel's Income Tax Protestors Page.* Jon Siegel, 31 Jan. 2007. Web. 3 Mar. 2010. <http://docs.law.gwu.edu/facweb/jsiegel/Personal/taxes/IncomeTax.htm>

Winston, E. "Zeitgeist, the Movie Debunked." *Conspiracy Science.* Edward L Winston, 1 Jan. 2008. Web. 2 Mar. 2010. <http://conspiracyscience.com/articles/zeitgeist/>

17. THE GEORGIA GUIDESTONES

Dubbed "America's Stonehenge", this granite monument in Georgia appeals to all sorts of conspiracy theorists.

On a hilltop in rural Elberton, Georgia stands an incongruous monument: 119 tons of imposing granite in five columns, six meters high, topped with a capstone. Not only are they geometrically arranged, they also offer three astronomical viewing ports marking the positions of the sun and stars, thus their nickname "America's Stonehenge". But unlike their English namesake, the Georgia Guidestones are not ancient; they are a recent emplacement. Their most significant trait is the controversial inscriptions, in eight languages. Some feel the messages in these inscriptions unite us; others feel they divide us. The monument refers to these ten mottos as the Guiding Thoughts, offered to the world by the anonymous builders of the monument. The ten Guiding Thoughts are:

MAINTAIN HUMANITY UNDER 500,000,000 IN PERPETUAL BALANCE WITH NATURE

GUIDE REPRODUCTION WISELY — IMPROVING FITNESS AND DIVERSITY

UNITE HUMANITY WITH A LIVING NEW LANGUAGE

RULE PASSION — FAITH — TRADITION AND ALL THINGS WITH TEMPERED REASON

PROTECT PEOPLE AND NATIONS WITH FAIR LAWS AND JUST COURTS

LET ALL NATIONS RULE INTERNALLY RESOLVING EXTERNAL DISPUTES IN A WORLD COURT

AVOID PETTY LAWS AND USELESS OFFICIALS

BALANCE PERSONAL RIGHTS WITH SOCIAL DUTIES

PRIZE TRUTH — BEAUTY — AND LOVE — SEEKING HARMONY WITH THE INFINITE

BE NOT A CANCER ON THE EARTH — LEAVE ROOM FOR NATURE — LEAVE ROOM FOR NATURE

Most of those are pretty harmless, and in fact a lot of people would probably agree that many of them are fine ideas. But the first three, recommending population control, eugenics, and a single world language, throw fuel on the fire of conspiracy theorists who fear an impending New World Order. If not for these first three controversial guides, the stones would probably be long forgotten.

The story of how the Guidestones came to be is just as mysterious as their meaning. The short version of it goes like this. A gentleman named Joe H. Fendley, Sr., was the president of one of the region's many granite companies, the Elberton Granite Finishing Company. According to the story, one day in June, 1979, he received a visit from a well dressed man who identified himself only with the pseudonym of Robert C. Christian. He told Fendley he represented "A small group of loyal Americans who believe in God." Christian showed Fendley the plans for his monument and asked for a price. Interestingly, Christian gave his specifications using the metric system, which was pretty rare in the United States in 1979. Fendley later said the price he gave was "six figures". Christian inquired where Fendley banked, and then left to go meet with the banker.

Thirty minutes later, Christian was received by Wyatt C. Martin, the president of the Granite City Bank. Christian expressed the need for secrecy on this project, including that the funds would be transferred to Martin from a number of different banks around the country. As a banker, Martin did require that Christian show him his true identification, which Christian did, only after Martin agreed to absolute secrecy in perpetuity. Martin agreed to be Christian's representative in managing the project going forward. Christian then left, corresponding only with Martin from that moment on, and often from different cities.

Fendley's company went to work. On Christian's behalf, Martin purchased fives acres of land atop the highest hill in the county for $5000 from farmer Wayne Mullenix, who subsequently acted as contractor laying the foundation for the monument. The six granite slabs were cut, drilled, lettered, and positioned according to the specifications, and unveiled on March 22, 1980. About 400 curious onlookers attended the opening.

In the decades since, the Georgia Guidestones have sat on their hilltop, gradually succumbing to vandalism, although a number of local volunteers have tried to keep them maintained. Other supporters have offered to install benches or pathways. Local churches are said to disapprove, but since the Guidestones have proven to be the most significant tourist attraction in Elberton, the town lets them be. The Sheriff has even installed security cameras to keep an eye on them.

The author at the Guidestones

The biggest question that everyone wants to know about the Guidestones is who built them? The stones themselves give one answer. A flat stone in the ground gives some information about the monument, and lists as its sponsors "A small group of

Americans who seek the age of reason." This answer, quite obviously, is unsatisfactory to the curious. According to Christian, this was by design: he once said "The group feels by having our identity remain secret, it will not distract from the monument and its meaning." I happen to think he was right on the money. If the monument was known to have been erected by a particular group, it would be easy to dismiss it as "Oh, just more of that nonsense from so-and-so." The lack of a source lets the Guidestones stand on their own, and the mystery keeps people interested.

The nomination of candidates is easily started. The most obvious name is that of Robert C. Christian himself: Was he actually the only person involved? Although Christian and Martin, the banker, corresponded quite regularly through the years and occasionally even met for dinner, Martin now says that he has not heard from him for some time and assumes he has passed away. Christian said he chose that particular pseudonym because he was a Christian. Some, most notably the conspiracy theorist Jay Weidner, have pointed out the similarities between the name R. C. Christian and that of the apocryphal founder of Rosicrucianism, Christian Rosenkreuz. But whoever Christian was, he was probably not an official representative of Rosicrucianism. Rosicrucian philosophies and mysticism are well documented and widely available, and they bear only minimal similarity to the Guiding Thoughts. L. Ron Hubbard has also been suggested as Christian's true identity, but this is an even worse fit.

Martin and Fendley are indeed both real people, which is something that you do need to check out in stories like this. The late Mr. Fendley served as mayor of Elberton from 1980 through 1987; and Wyatt Martin is still alive, in fact *Wired* magazine interviewed him in 2009. Could either Martin or Fendley have been the real sponsor, or even colluded as co-sponsors and made up the alleged R. C. Christian? It's a possibility. Nobody else ever saw or spoke with Christian. Martin showed *Wired's* reporter a box of his correspondence with

Christian, but would not allow the papers to be examined, citing his promise of secrecy. It's often noted that Fendley was a 32nd-degree Freemason, and several of his top workers who participated in the Guidestones were also Freemasons in his same lodge. While this may be intriguing to conspiracy theorists, it is not useful evidence of anything at all. Since the Georgia Guidestones were a private enterprise, there are no public records of its financing trail, except the property purchase, which has been fully disclosed. Thus, if Martin and Fendley were indeed responsible, there is no evidence for it, and little likelihood of any evidence ever coming to light. There is, in fact, no reason that I could find to doubt the story as they've told it.

The best argument I can think of to point the finger at Martin and Fendley is that they lived in Elberton, near the monument; and Elberton is an awfully strange place to erect a monument that you want to be seen by the whole world. Martin and Fendley, if they wanted to build such a monument, might well choose to do it locally. However, you could make this same argument wherever in the world the monument was built, citing whoever contracted its construction. This small fact is an inevitability of the circumstance, it's not evidence; and does not suggest Martin and Fendley as the sponsors.

Conspiracy theorists also point to the name of the type of granite used: Pyramid Blue. Clearly (to some), the pyramid imagery suggests Masonic motives. In fact, the name comes from the name of the quarry, Pyramid Quarries, Inc., also in Elberton. Why was this quarry chosen? Simply because Fendley happened to own it. And he didn't even name it: He purchased the company in the 1970's as a supplier for his granite finishing and pet memorial companies.

There's one more clue in the Georgia Guidestones that points conspiracy theorists to suspect an origin in some Zionist New World Order, and that's the eight languages in which the Guiding Thoughts are inscribed. You might expect the eight most common languages to have been used. You would be in error. The languages are English, Spanish, Swahili, Hindi, Hebrew, Arabic, Mandarin, and Russian. Missing from the top eight are Bengali, Portuguese, Japanese, and German. Why are Hebrew and Swahili included? They're not even in the world's top *fifty* languages. Even if you're only considering languages spoken in the United States, the top eight are still missing Cantonese, French, German, Tagalog, Vietnamese, and Italian. I'm not sure what the inclusion of Swahili tells us, but the inclusion of Hebrew suggests to conspiracy theorists that a Zionist New World Order is responsible for the Guidestones. This, of course, presumes the existence of a Zionist New World Order conspiracy, and it presumes that such a conspiracy would be usefully served by the erection of a random granite monument in the middle of nowhere in Georgia. Both presumptions strain credibility.

I dispelled my own hypothesis that the true sponsor was merely a fan of *2001: A Space Odyssey* with a quick calculation showing the main stones' proportions as 1×4×10, just off from Clarke's squares of 1, 2, and 3.

The facts of the Guidestones' history and construction is publicly available in great detail, in a 50-page book published by the Elberton Granite Finishing Company, called *The Georgia Guidestones*. It includes biographies of many of the principal workers and characters in the story, and dozens of photographs of the monument and its construction process. Anything you want to know about the Guidestones can be found in this book. Is the publication of this book evidence that Fendley was behind the whole thing? Not really, because there is another explanation that does not require such an assumption. The Guidestones were the most expensive project Fendley ever did, and by far the most famous. As such, they were the best possible advertisement for his company. What business would not leverage such a marketing opportunity? So, again, Fendley's promotion of the Guidestones were merely an inevitability of the circumstance.

In a world with free speech, there are going to be Georgia Guidestones, and all manner of similar exhibits. If you don't expect to find them outside of the existence of a New World Order conspiracy, you are being unrealistic. There's no proof that the Guidestones are *not* evidence of a conspiracy, but they're also exactly consistent with what we'd expect to find *without* such a conspiracy.

References & Further Reading

Bridges, B. *The Georgia Guidestones.* Elberton: Elberton Granite Finishing Company, Inc., 1980.

Curators. "The Georgia Guidestones (sculpture)." *Smithsonian Institution Research Information System.* Smithsonian Institution, 1 Nov. 1993. Web. 23 May. 2011. <http://siris-artinventories.si.edu/ipac20/ipac.jsp?&profile=all&source=~!siartinventories&uri=full=3100001~!333035~!0>

Lewis, M. Paul (ed). *Ethnologue: Languages of the World, Sixteenth Edition.* Dallas: SIL International, 2009.

May, L. "In one small town, the human condition is etched in stone." *Los Angeles Times.* 4 Sep. 1989, Volume 108.

Schemmel, W. *Insiders' Guide Off the Beaten Path Georgia.* New York: Globe Pequot, 2006. 205-206.

Sullivan, R. "American Stonehenge: Monumental Instructions for the Post-Apocalypse." *Wired.* 20 Apr. 2009, Volume 17, Number 5.

18. Cargo Cults

> *We point the skeptical eye at native religious groups in the South Pacific hoping to recreate WWII's influx of material goods.*

It was a beautiful day on Planet 4 of System 892 when Kirk, Spock, and McCoy materialized. McCoy took in the surroundings and expressed the thought so many of us have had: "Just once I'd like to be able to land someplace and say: 'Behold, I am the Archangel Gabriel.'" It sounds like a fun joke, but take it seriously for a moment. As the characters opined in another episode when they encountered a godlike being:

> To the simple shepherds and tribesmen of early Greece, creatures like that would have been gods... Especially if they had the power to alter their form at will and command great energy. In fact, they couldn't have been taken for anything else.

And now let's take these Star Trek references and see how they apply to real life; in particular, and in an extraordinary segue, to the tropical islands of the South Pacific. What happens when you mix native populations with modern visitors? In some cases, what's happened has been a curious religious phenomenon known as "cargo cults".

If you've heard of cargo cults before — and a lot of people have not — the version that you heard probably goes something like this. During WWII in the Pacific theater, Allied troops landed on islands throughout the South Pacific, bringing with them food, medicine, Jeeps, aircraft, housing, electricity, refrigeration, and all manner of modern wonders that the native populations had never seen before. But then the war ended and the troops went home, leaving just a few scraps behind. The natives, in a demonstration of Arthur C. Clarke's third law which states "Any sufficiently advanced technology is indistinguishable from magic," concluded that such a windfall must have come from the gods. They wanted this wealth of cargo to return. And so they did what seems logical from a stone-age perspective: they set about to recreate the conditions under which the gods and their cargo had come. They cleared paths in the jungle to resemble airfields. They wore scraps of military uniforms. They made "rifles" out of bamboo and marched as they had seen the soldiers march. And always they kept their eye on the sky, hopeful that the gods observed their preparations and would soon return with more cargo.

On some islands, particularly the New Hebrides (now called Vanuatu), these gods were personified in John Frum, an apocryphal American serviceman, according to the most popular version of the tale. John Frum is symbolized by a red cross, probably inspired by that painted on the sides of ambulances during the war. To this day, a surviving core of the John Frum Movement dresses in imitation WWII uniforms and celebrates February 15 as John Frum Day, in a plaza marked with a red cross and an American flag. They predict that on this day, John Frum will eventually return, bringing all the material goods of the modern world with him. In the words of one village elder:

> *John promised he'll bring planeloads and shiploads of cargo to us from America if we pray to him. Radios, TVs, trucks, boats, watches, iceboxes, medicine, Coca-Cola and many other wonderful things.*

The story of John Frum is sometimes erroneously confused with Tom Navy. Tom Navy was probably an actual person, possibly Tom Beatty of Mississippi, who served in the New Hebrides both as a missionary, and as a Navy Seabee during the war. Tom Navy is more of a beloved historical character associated with peace and service, whereas John Frum is regarded as an actual messiah who will bring wealth and prosperity.

The popular version of the John Frum story may seem a little whimsical. It's actually quite oversimplified and misstates the actual causes and motivations behind what happened. This particular cargo cult has deeper roots that have pulled directly on the heartstrings of much of the population. It goes all the way back to the early 18th century, long before anyone thought of World War II or American servicemen. At that time, the New Hebrides were an unusual type of colony called a condominium, jointly administered by both the British and the French. Among the early colonists were Scottish Presbyterian missionaries, who took a dim view of the uninhibited native lifestyle. On the island of Tanna beginning around 1900, at which time there was no meaningful colonial government, the missionaries imposed their own penal system upon the natives, a period called Tanna Law. Many of the traditional practices were banned, including ritual dancing, polygamy, swearing, and adultery. They also required observation of the Sabbath. But perhaps their most inflammatory prohibition was that of the traditional practice of drinking kava by the men. Those who violated these rules were convicted by the missionaries and sentenced to hard labor.

So it was a population in dire need of a savior to whom John Frum first appeared, and he did so in the 1930's. By most contemporary accounts, John Frum was a native named Manehivi who donned Western clothes; only in later versions of the story did he become an American serviceman. John Frum advocated a new lifestyle that was a curious mixture of having your cake and eating it too. He promised that if the people followed him, they could return to their traditional ways, but he would also reward them with all the material goods that the missionaries had brought. And so this is what the majority of the islanders did: The missionaries were suddenly ignored and found themselves vastly outnumbered by a population who took renewed interest in all their previous freedoms. Colonial authorities were summoned and leaders of the movement, including several chiefs, were arrested and imprisoned in 1941, introducing a new and culturally powerful element into the situation: martyrdom.

And then, an extraordinary thing happened. World War II descended upon the Pacific. The New Hebrides were flooded with Westerners. Food, medicine, Coca-Cola, and money were showered upon the natives. Many islanders were recruited as laborers and paid (relatively) lavishly. Life was rich with both traditional freedoms and material wealth. John Frum's promise had been miraculously fulfilled.

And so it's clear that the John Frum Movement has more to it than just a silly superstition that if you build something that looks like a dock out of bamboo, supply ships will come streaming in. That's how cargo cults are often portrayed, and it's really not a fair description. The people were going through genuine oppression, a man stepped up and promised freedom, and he delivered in spades. That actual fulfillment of prophecy, though it was merely a fortuitous coincidence, is still more than a lot of other religions can claim. So it does make a certain amount of sense that today's members of the John Frum Movement still look out to sea, and to the sky, waiting for their bounty. As one modern chief explained:

> *John was dressed in all white, like American Navy men, and it was then we knew John was an American. John said that when the war was over, he'd come to us in Tanna with ships and planes bringing much cargo, like the Americans had in Vila.*

Historians have not made much progress trying to find the origins of the name John Frum. One interesting explanation is that "frum" happens to be the pijin pronunciation of broom, as in sweeping the white people off the island. It's also likely that there was an actual person in the islands with a German last name of Fromme or Frumm, and Manehivi could have adopted his name. Another possibility is that it's a simple contraction of "John from America".

Cargo cults have appeared many, many times, and were not all centered around WWII. One of the earliest known cargo cults grew on the Madang Coast of Papua New Guinea, when the pioneering Russian anthropologist Nicholas Miklouho-

Maclay stayed there for some time in the 1870's, bringing with him gifts of fabric and steel tools. A hundred years later, a group formed on the island of New Hanover, and believed that if they could acquire American president Lyndon Johnson and install him as their king, cargo would come along with him. They rebelled against the Australian authorities, formed their own government, put together a budget, and offered to purchase Lyndon Johnson from the United States for $1,000. Their price was probably naïve, but just think what would have happened had Johnson accepted: Their plan probably would have worked better than they ever imagined.

The blending of Christianity with native superstitions sometimes caused some interesting problems. During WWII, some Australian groups grew concerned with what they saw as the sacrilegious inclusion of cargo cult principles with Jesus in Papua New Guinea. An educated New Guinean official named Yali, who had been on good terms with the missionaries, was employed by the Australians to travel around and try to dispel cargo cult mythology. After the war, Yali was rewarded with a trip to the Australian mainland, where he saw three things that greatly disturbed him, and caused him to rethink his work of the past few years. The first was the obvious wealth of the Australians compared to New Guinea. The second was a collection of sacred New Guinea artifacts on display at the Queensland Museum, which he began to suspect had been stolen by the Australians and resulted in their great accumulation of material goods. The third, and perhaps most influential, was exposure to the theory of evolution. This led Yali to conclude that the Australian missionaries, who had promoted the story of Adam and Eve, had been lying to him. Taken altogether, Yali reflected that he had been right to preach the separation of Christianity and cargo cults, but that he'd been on the wrong side.

And so while cargo cults may seem, at first glance, like quaint stone age ignorance, they're actually not entirely irrational. They're certainly naïve and based on a fallacious confusion of correlation and causation, but to give their believers some credit, they're doing their best to make sense of what they've been given. Where this belief system fails them, quite obviously, is that it replaces the need to work hard to achieve goals with the belief that faith will provide. This is the lesson that would best serve the believers, and it's the same lesson that missionaries and social workers should pay the most attention to. Rather than smiling at their funny little religion, or trying to replace it with another, we should instead give them the tools they need to create their own wealth of cargo.

References & Further Reading

Bonnemaison, J., Pénot-Demetry, J. *The Tree and the Canoe: History and Ethnogeography of Tanna.* Honolulu: University of Hawaii Press, 1994. 199-218.

Muller, K. "Tanna Awaits the Coming of John Frum." *National Geographic.* 1 May 1974, Volume 145, Number 5: 706-715.

Raffaele, P. "In John They Trust." *Smithsonian.* 1 Feb. 2006, Volume 36, Number 11: 70-77.

Rice, E. *John Frum He Come: Cargo Cults & Cargo Messiahs in the South Pacific.* Garden City: Dorrance & Co., 1974.

Rush, J., Anderson, A. *The Man with the Bird on His Head: The Amazing Fulfillment of a Mysterious Island Prophecy.* Seattle: YWAM Publishing, 2007.

Worsley, P. "Cargo Cults of Melanesia." *Scientific American.* 1 May 1959, Volume 200, Number 5: 117-128.

19. The Virgin of Guadalupe

Is the Virgin of Guadalupe a miraculous apparition, a dismissible religious icon, or does it have more importance?

In this chapter, we're going to travel back to the time of the Conquistadors, when Spanish soldiers marched through Aztec jungles and spread Catholicism to the New World. We're going to examine an object that is central to faith in Mexico: An image called the Virgin of Guadalupe.

The Virgin of Guadalupe is basically Mexico's version of the Shroud of Turin. Both are pieces of fabric, hundreds of years old, on which appears an image said to be miraculous. Both are considered sacred objects. But the Virgin of Guadalupe is a much more powerful icon to many Mexicans. There's hardly anywhere you can go in Mexico and not find a reproduction of the image. Its importance as a religious and cultural symbol cannot be understated, for it came from the very hands of The Most Holy Virgin Mary, Our Lady of Guadalupe, Queen of Mexico and Empress of the Americas.

A legend well known in Mexico tells how it came to be. In 1531, the Spanish had been occupying Mexico for about ten years. An indigenous peasant, Juan Diego, was walking in what's now Mexico City when he saw the glowing figure of a teenage girl on a hill called Tepeyac. She identified herself as the Virgin Mary, and asked him to build her a church on that spot. Diego recounted this to the Archbishop of Mexico, Juan de Zumárraga (1468-1548). Zumárraga was skeptical and told Diego to return and ask her to prove her identity with a miracle. Diego did return, and encountered the apparition again. She told him to climb to the top of the hill and pick some flowers to present to the Bishop. Although it was winter and no flowers should have been in bloom, Juan Diego found an abundance of flowers of a type he'd never seen before. The Virgin Mary bundled the flowers into Diego's cloak, woven from common cactus fiber and called a tilma. When Juan Diego presented the tilma to Zumárraga, the flowers fell out and he recognized them as Castilian roses, not found in Mexico; but more significantly, the tilma had been miraculously imprinted with a colorful image of the Virgin herself. This actual tilma, preserved since that date and showing the familiar image of the Virgin Mary with her head bowed and hands together in prayer, is the Virgin of Guadalupe. It remains perhaps the most sacred object in all of Mexico.

The story is best known from a manuscript written in the Aztecs' native language Nahuatl by the scholar Antonio Valeriano (1531-1605), the *Nican Mopohua*. By the European watermark on its paper, it's known to have been written sometime after 1556. This was widely published in a larger collection in 1649 by the lawyer Luis Laso de la Vega. Zumárraga and Juan Diego were both dead by the time Valeriano wrote it, so where did he get his information?

The *Nican Mopohua*

A red flag that a number of historians have put forth is that Bishop Zumárraga was a prolific writer. Yet, in not a single one of his known letters, is there any mention of Juan Diego, his miraculous apparition, the roses, or the cloak bearing the image, or any other element of the story in which Zumárraga was alleged to have played so prominent a role.

Not everyone agrees. In the 2000 book in Spanish, *Juan Diego, una Vida de Santidad que Marcó la Historia (A Life of Holiness that Made History)*, author Eduardo Chavez Sánchez gives, at some length, various quotations from letters by Zumárraga that he believes confirms the Juan Diego narrative. I found his list to be extraordinarily unconvincing, and I would honestly describe it as really desperate scraping of the bottom of the barrel to find a quote-minable quote. In fact, the only quote from Zumárraga I found that was remotely close was:

> *An Indian goes to Brother Toribio and all will be in praise of God.*

That sounds great because he mentions an Indian talking to a Catholic figure, but there's no mention of this Indian's name, no mention in the Juan Diego stories of a Brother Toribio (that I could find), and no elements of the Juan Diego story included in this single-sentence snippet. So unless some more of Zumárraga's writings come to light, I'm going to agree with the historians who say Zumárraga wrote nothing of these events, which casts doubt on his role in something that would have been of such great importance to him.

The name Juan Diego itself suggests that the story was a fictional invention. It basically translates as John Doe, a generic everyman, whose identity is unimportant. This doesn't prove anything, since there certainly were real people named Juan Diego, but it is an intriguing element.

It is the actual image of Mary itself that tells us the most about its true history. As every schoolchild knows, Hernán Cortés (1485-1547) was the Spanish Conquistador who overthrew the Aztec empire and placed much of Mexico under Spanish control in 1521. He was born in a region of Spain called Extremadura, and grew up to revere Our Lady of Guadalupe, a statue of a black version of the Virgin Mary, at the Santa María de Guadalupe monastery in Extremadura. This statue is credited with miraculously helping to expel the Moors from Spain in the Reconquista. Cortés brought reproductions of this European image of Mary with him when he went to the New World. Her dark skin resembled the Aztecs, and she became the perfect icon for the missionaries who followed Cortés to rally the natives into Christianity.

One such missionary was Fray Pedro de Gante (1480-1572), a Franciscan monk from Belgium (born Pieter van der Moere) who learned the Aztec language and created the first European-style school in Mexico, San Jose de los Naturales. One of his promising art students was a young Aztec man with the Christian name Marcos Cipac de Aquino, one of three known prolific Aztec artists of the period. In 1555, the newly arrived Archbishop of Mexico, Alonso de Montúfar (1489-1572), successor to the deceased Zumárraga, was looking to commission a portrait of the Virgin Mary, as a sort of teaching aide to help convert the Aztecs. Montúfar found the young artist Marcos at de Gante's school. And so, in 1555, the Aztec artist Marcos Cipac de Aquino painted a portrait of the Virgin Mary, with dark skin, with head slightly bowed and hands together in prayer, on a common cactus-fiber canvas. The painting was named the Virgin of Guadalupe according to the tradition Cortés brought from Spain. Although the Extremadura statue was not in this pose, the pose was still one of European tradition. The most often cited example of Mary in this exact pose is the painting *A Lady of Mercy*, attributed to Bonanat Zaortiga and on display at the National Art Museum of Catalunya, painted in the 1430's. Marcos followed more than a century of European tradition.

There was a pragmatic element to Montúfar's introduction of this painting and allowing it to be worshipped. Before the Conquistadors, Tepeyac was home to an Aztec temple, built to honor the Aztecs' own virgin goddess, Tonantzin. So rather than replacing the Aztec goddess, Montúfar's plan was simply to introduce Mary by giving Tonantzin a name and a face (recall that Marcos had painted the Virgin with dark skin). This process of using an existing belief system to graft on a new one has been called syncretism. Understandably, this exploitation of a pagan idol caused discomfort among some of the Franciscan priests, while many of the Dominicans welcomed the way it helped baptize 8,000,000 Aztecs.

The primary corroborating documentation of Marcos' painting is a report from the Church in 1556, when this growing disagreement between the Franciscans and the Dominicans prompted an investigation into the origins of the tilma. Two of the Franciscans submitted sworn statements in which they expressed their concern that worshipping the tilma was leading the Aztecs to return to their traditional pagan ways. One described the image as "a painting that the Indian painter Marcos had done" while another said it was "painted yesteryear by an Indian". Appearing on the side of the Dominicans, who favored allowing the Aztecs to worship the image, was Bishop Montúfar himself. As a result, the construction of a much larger church was authorized at Tepeyac, in which the tilma was mounted and displayed.

Significantly, the 1556 report is the most extensive documentation concerning the Virgin tilma of its century, and it makes no mention whatsoever of Juan Diego, the miraculous appearance of the image, or any other element from the legend. If the miracle story did exist at that time, it seems inconceivable that it could have been omitted from this report. This strongly supports the suggestion that the Juan Diego legend had not yet been conceived. It also supports that Valeriano's *Nican Mopohua* was written later.

The legend did get its first boost of testable evidence in 1995, which (in a case of suspiciously fortuitous timing) was after Juan Diego's beatification in 1990, while there was still debate over whether he should be canonized (he ultimately was, in 2002). A Spanish Jesuit named Javier Escalada produced a deerskin that pictorially depicted the Juan Diego legend and has become known as the Codex Escalada. The Codex also mentioned several historical people, and even bore the signature of a Franciscan historian, Bernardino de Sahagún (1499-1590), dated 1548. Basically, it was the Perfect Storm of tailor-made evidence proving that the Juan Diego legend was the accepted history at the time. A little too tailor-made though; no serious historians have supported its authenticity. The best analysis I've found is by Alberto Peralta of the Proyecto Guadalupe project. Based on its dubious unveiling, numerous inconsistencies, and other factors, Peralta concludes that it's impossible for the document to be authentic.

If the Virgin tilma is indeed a painting, and not a miraculously produced image, then it should be a simple matter to determine that scientifically. There are obvious signs that are hard to argue with, notably that the paint is flaking along a vertical seam in the fabric. But a truly scientific examination involving sampling of the material has not been permitted. The most notable examination was a three hour infrared photographic session by Philip Callahan in 1981, who did note multiple layers of paint covering changes to the hands and crown, but came away with more questions than answers. Callahan found, for example, that most of the entire painting seemed to have been done with a single brush stroke. He recommended a series of more tests, but the only one allowed by the Church was a spectrophotometric examination done by Donald Lynn from the Jet Propulsion Laboratory. The only result released of his examination was that "nothing unusual" was found.

Much has been made of the claim that figures can be seen reflected in Mary's eyes, with some even identifying these figures as Zumárraga or Juan Diego or other characters from the legend. The Church even went so far in 1956 as to have two ophthalmologists examine the eyes under 2500× magnification. They reported a whole group of figures, including both Aztecs and Franciscans. Why ophthalmologists should be better qualified to identify Aztecs and Franciscans in random blobs of pigment has not been convincingly argued. Photos taken by another ophthalmologist in 1979 have been released, and it's quite obvious that it's simply random noise. I see a dozen or so speckles; if you want to make them into Aztecs, Franciscans, bananas, or Bozo the Clown, then you'll probably also be great at spotting dozens of Bigfoots hiding in any given photograph of a forest.

The Virgin of Guadalupe is yet more one mythical story whose believers are missing out on true facts that are actually more respectful and confer more credit upon them than the myth. The image on the Virgin tilma was painted by a native Aztec artist; and the painting had not only an important role in Mexico's early history as a nation, but also a staggering impact upon its culture ever since. Mexicans with Aztec heritage should take pride in the fact that their original culture, specifically the goddess Tonantzin, was a key ingredient in the spread of modern Catholicism. The Juan Diego myth takes that away, and whitewashes part of Mexican history clean of any Aztec influence. That's a disservice to one of humanity's greatest ancient civilizations, and it's a disservice to history.

When we see the Virgin of Guadalupe image today, most people react in one of two ways: They worship it as a miraculous apparition, or they dismiss it as someone else's religious icon. Both reactions miss the much richer true history. The Virgin of Guadalupe stands not only as an invaluable work of ancient art (possibly the most popular piece of art ever created), but also as a reminder of how the conquest of Mexico was truly accomplished: Not only its military conquest, but one of history's greatest religious conversions as well.

REFERENCES & FURTHER READING

Acosta, M. "Juan Diego: The Saint That Never Was." *Free Inquiry.* 1 Apr. 2003, Volume 23, Number 2.

Nickell, J., Fischer, J. "The Image of Guadalupe: A folkloristic and iconographic investigation." *Skeptical Inquirer.* 1 Apr. 1985, Volume 9, Number 3: 243-255.

Olimon, M. *La Búsqueda de Juan Diego.* Mexico City: Plaza & Janes, 2002.

Peralta, A. "El Códice 1548." *Proyecto Guadalupe.* ProyectoGuadalupe.com, 19 Dec. 2001. Web. 5 Apr. 2010. <http://www.proyectoguadalupe.com/apl_1548.html>

Sanchez, E. *Juan Diego, una vida de santidad que marcó la historia.* Mexico City: Editorial Porrúa, 2002.

Smith, J. *The Image of Guadalupe: Myth or Miracle?* Garden City: Doubleday, 1983.

20. The Mystery of Pumapunku

Were the stone structures at Pumapunku truly so advanced that the ancient Tiwanakans could not have made them unassisted?

In this chapter, we're going to climb high into the Andes and take a look at an ancient structure that has been cloaked with as much pop-culture mystery as just about any other on Earth: Pumapunku, a stone structure that's part of the larger Tiwanaku. Pumapunku, which translates to the Doorway of the Puma, is best known for its massive stones and for the extraordinary precision of their cutting and placement. It's one of those places where you've heard, probably many times, that the stones are so closely fitted that a knife blade cannot be inserted between them. Due to these features, Pumapunku is often cited as evidence that Earth was visited by aliens, Atlanteans, or some other mythical people who are presumably better at stonemasonry than humans.

Tiwanaku is in Bolivia, up in the Titicaca Basin, about 10 kilometers away from the great Lake Titicaca. The Titicaca Basin is high; 3,800 meters (12,500 feet) above sea level. Half is in Peru and half is in Bolivia, and right on the border sits Lake Titicaca. It's in a vast region of the Andes Mountains called the Altiplano, or "high plain", the largest such plain outside of the Himalayas. The Tiwanaku Culture predated the Inca, and their history is known largely from archaeology, since they had no written language that we know of. The earliest evidence of habitation dates from around 400 BC, but it wasn't until about 500 AD that the Tiwanaku Culture truly developed. At its peak, 400,000 people lived in and around the Tiwanaku site, centering around Pumapunku and other important structures. Trade and farming flourished. Farming was done on raised fields with irrigation systems in between them. Decades of drought struck around 1000 AD, and the city of Tiwanaku was abandoned, and its people and culture dissolved into the surrounding mountains. Five centuries later, the Inca Culture developed.

So within the context of Tiwanaku, Pumapunku does not leap out as extraordinary. However it does differ from the other structures at Tiwanaku, in that many of the blocks are shaped into highly complex geometries. There is a row of H-shaped blocks, for example, that have approximately 80 faces on them; and all match each other with great precision. Pumapunku's stones suggest prefabrication, which is not found at the other Tiwanaku sites. In addition, some of the stones were held together with copper fasteners, some of which were cold hammered into shape, and others that were poured into place molten.

Complex shapes at Pumapunku

Due to the complexity and regularity of many of Pumapunku's stone forms, a number of authors have suggested that they're not stones at all, but rather concrete that was cast into forms. We don't have any record that such technology was known to pre-Incan cultures, but that doesn't prove it wasn't. What can be proven, and proven quite easily, is that there is no concrete at Pumapunku or anywhere else in Tiwanaku. Contrary to the suppositions of paranormalists, modern geologists are, in fact, quite able to discern rock from concrete. Petrographic and chemical analyses are relatively trivial to carry out, and even allowed us to determine exactly where the rocks were quarried. Pumapunku's large blocks are a common red sandstone that was quarried about 10 kilometers away. Many of the smaller stones, including the most ornamental and some of the facing stones, are of igneous andesite and came from a quarry on the shore of Lake Titicaca, about 90 kilometers away. These smaller stones may have been brought across the lake by reed boat, then dragged overland the remaining 10 kilometers.

Much is often made of the vast size and weight of the Pumapunku stones, with paranormal web sites routinely listing them as up to 440 tons. Pumapunku does indeed contain the largest single stone found at any of the Tiwanaku sites, and it's part of its Plataforma Lítica, or stone platform. The accepted estimate of this piece of red sandstone's weight is 131 metric tons, equal to 144 US tons. The second largest block is only 85 metric tons, and the rest go down sharply from there. The vast majority of the building material at Pumapunku consists of relatively small and easily handled stones, although many of the most famous are megalithic. The absurd numbers like 440 tons come from much earlier estimates, and have long since been corrected.

We do not claim to know how the heavy lifting and exquisite masonry was accomplished at Pumapunku, but that's a far cry from saying we believe the Tiwanaku were incapable of it. We simply don't have a record of what tools and techniques they used. All around the world are examples of stonemasonry from the period that is equally impressive. The Greek Parthenon, for example, was built a thousand years before Pumapunku, and yet nobody invokes aliens as the only explanation for its great beauty and decorative detailing that more than rivals Pumapunku's angles and cuts. At about the same time, the Persians constructed Persepolis with its superlative Palace of Darius, featuring details that are highly comparable to Pumapunku. Stonemasons in India cut the Udayagiri Caves with megalithic doorways that are very similar to those in Pumapunku. The Tiwanaku did magnificent work, but by no means was it inexplicably superior to what can be found throughout the ancient world. It is unnecessary to invoke aliens to explain the structures.

Curiously, if you do an Internet search for Pumapunku, you'll find it almost universally, and quite casually, referred to as a "port". At least, this is what it's called on the paranormal web pages, which make up the overwhelming bulk of Pumapunku information on the Internet. In fact, it's not a port, and it never was a port. To anyone doing even the most basic research or visiting the area, it's a fairly bizarre assertion, considering that Pumapunku was in the middle of a vast farming nation of 400,000 people. Nor are Tiwanaku's structures in locations where they could serve as a port. Pumapunku is just one of several stepped platform constructions that have been excavated at Tiwanaku. The others include Akapana, Akapana East, Kalasasaya, Putuni, and the Semi-Subterranean Temple. If you look at them from above, they're simply squarish enclosures scattered about the area. If you imagine water filling the region — let's pretend just high enough to cover the ground but not the enclosures themselves — then each of these "ports" would be an island unto itself, amid a sea of knee-deep water too shallow to be navigable. (That is, except for the Semi-Subterranean Temple, which being recessed into the ground, would have been underwater.)

But even that imaginary scenario presumes that the lake could ever reach Tiwanaku. It can't. The Altiplano is a vast sloping plain, and the point at which Lake Titicaca spills off the edge into its sole outlet, the Río Desaguadero, is about 30 meters *below* the elevation of Tiwanaku. Has this always been the case? At least since the last ice age, yes. Because the sediment at the lake's bottom has been accumulating for some 25,000 years, it's one of the best places to get data about Earth's climate history, and so it's been extensively studied. The paleohydrology of Lake Titicaca is thoroughly known. Currently, the water is at its overflow level. This level has fluctuated about 5 meters in the past century. During the past 4,000 years, it has dropped as much as 20 meters during drought periods. The maximum it's ever been is about 7 meters above overflow level, which would still locate the shore many kilometers away from Tiwanaku's suburbs and farms.

Most of these same paranormal sources that refer to Pumapunku as a port also state that the ancient shoreline is still visible along the surrounding hills, albeit tilted at a strange angle. Ancient lake levels are often visible in such a way — they're quite prominent throughout Death Valley where I often visit, for example. But it makes no sense for Lake Titicaca. The lake would spill off the edge of the plain before it could get as high as Tiwanaku; and there's certainly been no tectonic activity in that time that could have tilted the hills, or mysteriously tilted the hills yet left the Tiwanaku structures level.

Finally, one other feature at Pumapunku is said to have the archaeologists baffled: Carved figures, said to represent an elephant relative called a Cuvieronius, and a hoofed mammal called a toxodon. These both went extinct in the region around 15,000 years ago, and so some paranormalists have dated Pumapunku to 15,000 years, apparently based on this alone. When you hear that an elephant is carved there, it certainly does give you pause, because an elephant is hard to mistake. However, when you look at a picture of what's claimed to be the elephant, this becomes less surprising. Tiwanaku art was highly stylized, much like what we're accustomed to seeing from the Mayans or the Aztecs. It's actually the heads of two crested Andean Condors facing each other neck to neck, and their necks and crests constitute what some have compared to the tusks and ears of an elephant's face. The image of the toxodon is known only from rough sketches of a sculpture discovered in 1934, and so it's a drawing of indirect evidence of an artist's interpretation of an unknown subject. It looks to me like a generic quadruped. Pig, dog, rat, toxodon, name it.

So once again, we have an accomplishment by ancient craftsmen whom some paranormalists have attempted to discredit by attributing their work to aliens. This is not only irrational, it's a non-sequitur conclusion to draw from the observations. Most people don't know how to intricately cut stones because those are skills we haven't needed for a long time — we've had easier ways to make better structures for a long time. But this argument from ignorance — that just because we don't know how to do it, nobody else could have figured it out either — is an insufficient explanation. Simply say that you don't know, instead of invoking aliens. This is not only the truth, it accurately represents the findings of science so far; and perhaps most importantly, it leaves the credit for this wonderful contribution to humanity where it belongs: with the Tiwanaku themselves.

References & Further Reading

Abbot., M., Binford., M., Brenner, M., Kelts, K. "A 3500 14C yr High-Resolution Record of Water-Level Changes in Lake Titicaca, Bolivia/Peru." *Quaternary Research.* 1 Jan. 1997, Volume 47: 169-180.

Baker, P., et. al. "The History of South American Tropical Precipitation for the Past 25,000 Years." *Science.* 26 Jan. 2001, Volume 291.

Janusek, J. *Identity and Power in the Ancient Andes: Tiwanaku Cities Through Time.* New York: Routledge, 2004. 133-137.

Ponce Sanginés, C., Terrazas, G. "Acerca De La Procedencia Del Material Lítico De Los Monumentos De Tiwanaku." *Academia Nacional de Ciencias de Bolivia.* 1 Jan. 1970, Number 21.

Pratt, D. "Lost Civilizations of the Andes." *Exploring Theosophy: The Synthesis of Science, Religion and Philosophy.* David Pratt, 1 Jan. 2010. Web. 12 Apr. 2010. <http://davidpratt.info/andes2.htm>

Young-Sánchez, M. *Tiwanaku: Ancestors of the Inca.* Denver: Denver Art Museum, 2004. 32-34.

21. Therapeutic Touch

Therapeutic touch is a healing method that presumes the existence of a huge magnetic field around the human body. Is there really such a field?

In this chapter, we're going to point the skeptical eye at a healing modality that's non-invasive, requires no drugs, produces no unpleasant side effects, and is taught in nursing schools everywhere: therapeutic touch. As it does not actually involve any touching, it's also called distance healing, or non-contact therapeutic touch. The method is premised on the presumed existence of an "energy field" which practitioners believe extends around the human body, and which they believe they can see or touch, many claiming it's tactile and they can actually feel it. This "energy field" is then manipulated by waving the hands over the patient's body. Based on a number of varying explanations, this action is said to promote faster healing of wounds, reduce pain or anxiety, or promote wellness in some other way.

Therapeutic touch was developed in 1972, a collaboration between a self-described clairvoyant and a nursing professor. Dora Kunz was a member of the Theosophy Society in America, and soon to become its president. Theosophy is a form of religious mysticism which Kunz had studied under the Hungarian psychic healer Oskar Estabany. Dolores Krieger, who taught nursing at New York University, became engaged with Kunz's metaphysical approach to healing, and codified therapeutic touch into the system that it is today.

What sets therapeutic touch apart from most other alternative treatments is its level of acceptance within the medical community. Although it's made no inroads in medical schools or with doctors, it is taught in a surprising number of nursing schools and in continuing education for nurses. This apparent rubber stamp of legitimacy has convinced many that there's something to it. The University of Maryland Medical Center is one hospital that offers therapeutic touch, and on its web site, attempts to explain how they believe it works (since removed from their site). They offer two explanations, erroneously describing them as theories:

> *One theory is that the actual pain associated with a physically or emotionally painful experience (such as infection, injury, or a difficult relationship) remains in the body's cells. The pain stored in the cells is disruptive, and prevents some cells from working properly with other cells in the body. This results in disease. Practitioners believe therapeutic touch promotes health by restoring communication between cells.*

I read this five times and still couldn't make any sense of it. Pain remains in the cells? Pain is transmitted by nerves, and does not exist in the body outside of nerves, so right off the bat this makes no medical sense in even a rudimentary way. Then they say this is the cause of disease. No, not exactly. We have very good theories of various diseases and almost all are well understood, and nowhere in the medical texts do we find "the pain of difficult relationships remaining the body's cells". Finally, wrapping this up with the assertion that "therapeutic touch promotes health by restoring communication between cells" is not, in fact, an explanation of how anything works. It is a nonspecific claim loosely bolstered by a few vaguely medical-sounding words.

> *The other theory is based on the principles of quantum physics. As blood, which contains iron, circulates in our bodies an electromagnetic field is produced. According to this theory, at one time we could all easily see this field (called an aura), but now only certain individuals, such as those who practice therapeutic touch, develop this ability.*

Ah yes, the tired old "appeal to quantum physics". Mention that phrase, and immediately you sound smart. Well, not so much in this case: circulating iron in the blood creating a magnetic field is not an example of quantum physics. It's also known to be untrue: the tiny amount of iron in our hemoglobin, each molecule of which has a jumbled pole pointing in a different direction, neither creates a magnetic field, nor is there any electromagnetic theory which predicts that it might. And, there has never been a sound theory that all humans could at one time see magnetic fields, and no example of anyone ever seeing one. If anyone could, they would be essentially blinded all day by the overwhelming influence of the Earth's magnetic field, which is much stronger than the tiny ones we encounter all day long.

Dolores Krieger, one of the co-founders of therapeutic touch, offers yet another completely different explanation for how it works. In 1975, she wrote in the *American Journal of Nursing* that this manipulation of the body's magnetic field raises the body's hemoglobin levels. This statement packs about as many implausible falsehoods into one sentence as I've ever seen. Placing a hand into a magnetic field does not alter that magnetic field. A magnetic field strong enough to extend several inches out from the human body would have to be immensely strong, and we wouldn't be able to wear metal jewelry or get out of our car, and we'd be sticking to each other all day. There is no explanation for how or why any magnetic effects would create hemoglobin, and no report of such a thing ever happening from magnetic resonance imaging or other applications of real magnetic fields to the human body. Finally, raising hemoglobin levels may be an effective therapy for some conditions, but not for pain, which is the primary use of therapeutic touch. One has to marvel at how Krieger, an experienced medical professional, could have strung together such an outlandish chain of nonsense and promoted it as sound science.

The most famous example of therapeutic touch's basic premise being put to the test was in 1996, when a 9-year-old girl, Emily Rosa, ran an experiment for her fourth grade science fair. Her parents, who were both critical of therapeutic touch, helped her devise a simple protocol. She visited 15 practitioners who claimed to be able to detect the presence of a "human energy field", and had them stick both hands through a screen. She then held her own hand directly above one of the practitioner's hands, selected randomly, and the practitioner stated each time which hand detected her field. They failed to guess better than random chance, indicating that practitioners cannot, in fact, do what they claim. The following year, the TV show *Scientific American Frontiers* filmed her repeating the test, with the same results. The year after that, Dr. Stephen Barrett contacted the Rosas and helped them write it up and had the study published in the *Journal of the American Medical Association*. This gave young Emily, then 11, the world record for the youngest person to have a scientific paper published in a peer-reviewed journal.

As you can imagine, the therapeutic touch community responded harshly. Healing Touch International, which was at the time the leading self-certification institute for practitioners, published an official response to the Journal's article with several key criticisms:

> *The published study does not test any critical variables related to therapeutic touch. The ability to sense the energy field of another is simply not a requirement of a TT practitioner.*

This statement that "the ability to sense the energy field of another is simply not a requirement" is staggering. The entire practice is founded on manipulation of this energy field, which is described by practitioners as "tingling, pulling, throbbing, hot, cold, spongy, and tactile as taffy". They've now moved the goalposts. They've gone from sensing and manipulating this alleged energy field, to stating that it's not necessary to actually sense anything. How are you supposed to manipulate something you can't detect?

Healing Touch International's response to Rosa's study continues:

> The study design was not representative of a therapeutic touch session. It was set up more as a parlor game. Therapeutic touch studies using people with health problems would most likely demonstrate positive effects.

They are correct that Rosa's study was not to test the efficacy of therapeutic touch on sick patients, it was about something else. It showed that therapeutic touch's fundamental concept is nonexistent. However, the studies they ask for do already exist. The authors of Rosa's paper included a literature analysis of 853 such studies. They concluded:

> There is not a sufficient body of data, both in quality and quantity, to establish TT as a unique and efficacious healing modality... With little clinical or quantitative research to support the practice of TT, proponents have shifted to qualitative research, which merely compiles anecdotes.

Since the publication of Rosa's study, the healing touch industry now publishes only research that considers the improvement reported by patients, without regard to the method used. Most therapeutic touch web sites cite a small number of studies where a small effect was shown, but ignore the much larger number of larger, better controlled studies that show no effect. This suggests deliberate deception on their part; it's impossible to do an honest search of the literature without happening to notice the saturation of large studies showing no results.

The next point in their response:

> The child conducting the study was not neutral about therapeutic touch, and therefore could have affected the results. The studies of Dan O'Leary at SUNY-Stony Brook have shown the "experimenter bias" effect, in which the beliefs of the ex-

perimenter tend to be confirmed. Controlled experiments look like they cannot possibly be affected by the experimenter's beliefs, but O'Leary showed that in the real world, they are.

The only thing I can think of to say to this is "pot-kettle-black". Emily Rosa did employ good controls, of which the same cannot be said for the majority of the studies that find therapeutic touch to have a beneficial effect. It's fine to point out that she was a child, and it's fine to say that someone else once found that some studies are weakened by experimenter bias; but if you want to poke holes in Rosa's study, you have to actually find some. You can't merely insinuate that others have found holes in other studies, thus the child's results are automatically suspect. The editor of the Journal said "Age doesn't matter. All we care about is good science. This was good science."

We could go on all day like this, but it's pointless. The only thing that matters when we evaluate a new treatment is whether it works. The therapeutic touch proponents will quote studies that show that it does work, usually Wirth's 1990 study, "The Effect of Non-Contact Therapeutic Touch on the Healing Rate of Full Thickness Dermal Wounds". But they fail to note that an effect was found in only two of Wirth's five trials, and *Wirth's own conclusion* was "The overall results of the series are inconclusive in establishing the efficacy of the treatment interventions examined." The best state of our experimental knowledge is that therapeutic touch does not work. It certainly has no plausible foundation, and no physiological reason to suspect that the body's healing mechanisms are dependent upon some outside person waving their hands around. Really this is one for the kindergarten files, and responsible nurses should look instead to treatments that have credible hope of doing some good.

References & Further Reading

Butler, K. "Therapeutic Touch: Are We Being Touched?" *The Skeptic.* 1 Jul. 2000, Volume 20, Number 3: 8-9.

Courcey, K. "Further Notes on Therapeutic Touch." *Quackwatch.* Stephen Barrett, MD, 14 Dec. 2002. Web. 16 Apr. 2010. <http://www.quackwatch.org/01QuackeryRelatedTopics/tt2.html>

Ehrlich, S. "Therapeutic Touch." *Medical Reference, Complementary Medicine.* University of Maryland Medical Center, 24 Aug. 2009. Web. 17 Apr. 2010. <http://www.umm.edu/altmed/articles/therapeutic-touch-000362.htm>

Krieger, D. *Therapeutic Touch Inner Workbook.* Santa Fe: Bear & Company Publishing, 1996.

Rosa, L., Rosa, E., Sarner, L, Barrett, S. "A Close Look at Therapeutic Touch." *Journal of the American Medical Association.* 1 Apr. 1998, Vol. 279 No. 13: 1005-1010.

Scheiber, B., Selby, C. *Therapeutic Touch.* Buffalo: Prometheus Books, 2000.

22. Mengele's Boys from Brazil

Is the unusually high twin rate of a town in Brazil due to continued medical experiments by Josef Mengele, the infamous Nazi Angel of Death?

Some may remember Franklin J. Schaffner's 1978 movie *The Boys from Brazil,* featuring the infamous Nazi death camp doctor Josef Mengele, nicknamed the Angel of Death. In the film, Mengele, played by Gregory Peck, had been living in Paraguay and spent 20 years producing clones of Adolf Hitler. These black haired, blue eyed "boys from Brazil" were seeded throughout the world, in the hope that one may grow to take Hitler's place. Laurence Olivier played a Nazi hunter who tracked down Mengele to a farm in Pennsylvania, where they engaged in the film's climactic duel to the death. The movie was nominated for three Academy Awards, and while many realized it was fiction, some no doubt assumed it was based on a reasonable account of actual events. How much truth is there to the story?

Complicating this question is the fact that enough building blocks from Mengele's true life story can be arranged to create a reasonable foundation for this. Jorge Camarasa is an Argentinian historian who has specialized in the exodus of Germans and other Europeans to South America before, during, and after the second World War. In 2008 he wrote a book called *Mengele: The Angel of Death in South America* in which he noted that a town in Brazil, Cândido Godói, has a rate of twin births that is ten times higher than normal; and he attributes this to a series of visits made there by Mengele under an assumed name during the 1960's. Did Josef Mengele continue his human experimentation in Brazil after the war?

Mengele arrived in Auschwitz as a capable young medical doctor and researcher, only 32 years old. Before joining the military, he'd worked at the Institute for Hereditary Biology and Racial Hygiene, and developed a particular interest in multiplying the Aryan race through multiple births. Auschwitz provided him the perfect opportunity: A large experimental population over which he had complete control. His first duty was to find all the twins, and he earned his nickname "the Angel of Death" as one of the selectors who decided the fate of each new arrival to the camp. He instructed his assistants to especially look for twins, dwarves, or anyone with any kind of physical irregularity.

Mengele wanted to know what caused Aryan traits, and identical twins afforded him the opportunity to experiment on one while keeping the other as a control. He tested hypotheses about the causes of eye color and other racial traits. Dissections and comparative autopsies are believed to have been a large part of this. Much of what actually happened will never be known, since the Nazis destroyed all of Mengele's records, leaving us with only survivor testimony to go on. The stories they tell are an appalling patchwork of surgeries and butchery, often with depraved sexual overtones, almost always without anesthesia, and arguably constitute the darkest moment in all of human history.

Contrary to popular retellings, Mengele was not doing any kind of research having to do with genetics, and the reason we know this for a certainty is that the science of genetics had not yet been invented. DNA was not yet known until the 1950's. So while Mengele may have had an interest in heredity, there is no way he could have experimented with genes or manipulated twin heredity other than the trial and error method of selective breeding. I've not found any record that he attempted this, and it wouldn't have made much sense anyway because World War II didn't last long enough. In fact, from a study of Mengele's actual experiments at Auschwitz, I didn't find a strong reason to even suggest Mengele as a possible factor in South American twin populations.

At the war's end, Mengele did like most German soldiers and escaped west in hopes of surrendering to Americans rather than to Russians. He was successful, and giving a fake name, was released and lived peacefully in Bavaria for several years, working as a farmhand, careful not to divulge that he was a doctor. Mengele knew that he was a wanted war criminal, and in 1949 he made his way to Buenos Aires, Argentina via the German underground, and joined other former Nazis in relative safety. At first he continued working as a laborer, but after a few years felt safe enough to resume medical practice, and worked as an abortionist. After more than a decade, Israeli agents began stepping up their game, and Mengele's friend Adolf Eichmann (known as "the architect of the Holocaust") was captured. Mengele again faded into the woodwork. He spent a few years in Paraguay, then later moved to Brazil, relocating from town to town until he finally died in 1979.

Josef Mengele, 1956

Though reports often say Mengele was "posing" as a veterinarian in Brazil, he was in fact earning a living as a veterinarian. Farm animals were an important commodity, and Mengele had the means and knowledge to treat them, which included artificial insemination. He also worked as a farm manager for a time. I found no records that he ever worked as a doctor in Brazil, and certainly no reports that he ever did any further human experiments. His life in Brazil is fairly well documented, and at no time were his circumstances such that he could have been continuing his Auschwitz work in secret. Histories of some of the South American Nazis and those pursuing them read like modern spy novels, and there were only a few times when Mengele had any chance to settle down for as much as a year or two.

According to Camarasa, it was during one of these periods when Mengele lived in Colonias Unidas, Paraguay. Camarasa cites the testimony of several witnesses who recall that in the 1960's, Mengele made several short trips across the border into Cândido Godói, which like many towns in the region consisted almost entirely of German expatriates. Camarasa speculates that these are the times in which he performed his twin experiments. DNA had been discovered by then, but the science was in its infancy, and was being studied largely in the United States and Europe. Practical gene manipulation was not yet being done. Mengele had scarce opportunity to learn about it in the backcountry of Brazil. By that time, his medical knowledge was almost 20 years outdated.

Corroborating evidence for these alleged visits seems hard to come by. Even if Mengele did visit Cândido Godói, even Camarasa himself offers nothing firmer than speculation that he ever did any experimentation. Nobody in Cândido Godói reports ever having met or been treated by Mengele, and there is no record of a laboratory or victims.

So where does this leave us? We have no proof that Mengele performed twin experiments in Brazil, and also no proof that he didn't. When setting out to solve a mysterious event, first find out whether that mysterious event ever actually happened. Let's look and see if there actually was a five-fold or ten-fold explosion in the twin birth rate in Cândido Godói in the 1960's. When we do this, we quickly learn two things:

- It turns out that Cândido Godói experienced no notable change in the twin birth rate in the 1960's. According to the town's baptismal records, twins had been unusually common in the town since at least the 1920's, decades before Mengele's arrival. Twin births are still common in the town today, decades after Mengele's death. Even if Mengele had developed some ovulation induction drug, it would have affected only that generation; he had no knowledge or ability to modify genetic code, which would have been necessary to pass the trait to future generations.

- Cândido Godói's twin rate is very high, but not extraordinarily high compared to similar towns in the region. It turns out that many such communities, not only in South America but worldwide, consisting of small, isolated populations, often expatriates, have high twin rates. In particular, isolated villages in Nigeria and Romania have similar histories and similar twin rates.

So it appears that we do have a bit of a mystery to solve, but it's not the one we started with. Whether Josef Mengele ever visited these twin towns or not, it's known that it had no effect on their twin rate, and thus it would be illogical to introduce him as a possible explanation.

A better explanation for the high twin rate lies in established genetics, where we have something we call the "founder effect". When a community is founded by a small number of individuals, particular genetic traits can become common among their descendants. This is observed all over the world; the best example being the mitochondrial Eve. While she was not the only person alive at the time, say 200,000 years ago, her descendants intermarried with enough other people that eventually everyone in the small population was related to her, and their descendants eventually colonized the whole world.

The Amish population in the United States exhibits a founder effect where polydactyly (extra fingers and toes) was inherited by the community from one of the original founders who had the gene, and that gene is expressed more in the Amish community than in the United States at large. One island in Micronesia was decimated by a hurricane in 1775, leaving only twenty survivors, one with an extremely rare form of colorblindness. Today their rate of that type of colorblindness is 1,500 times higher than average.

As it happens, many of the founders of Cândido Godói and the many other German towns in Brazil and Paraguay came from a region of western Germany called Hunsrück, which has a higher than average twin rate. When Cândido Godói was first settled, the founding families happened to include seventeen pairs of twins. With such a high expression of the twin gene in the founding population, and being socially isolated within Brazil, it was a virtual certainty that Cândido Godói would become one of the Twin Capitols of the World... no need to introduce Josef Mengele or other external influences.

Paolo Sauthier, the historian at Cândido Godói's small museum, perhaps said it best: "People who are speculating about Mengele are doing so to sell books." Although the sensational explanation can often be the most compelling, it's usually not the true one. Dig enough to find the facts, and you'll almost always learn something new and true; and since it's real, you can apply it elsewhere. It's the value of knowledge.

REFERENCES & FURTHER READING

Camarasa, J. Mengele: *El Angel de la Muerte en Sudamerica*. Mexico City: Norma Libros, 2008.

Editors. "Nazi Mystery: Twins from Brazil." *Explorer*. National Geographic Society, 25 Nov. 2009. Web. 30 Apr. 2010. <http://channel.nationalgeographic.com/series/explorer/4087/facts>

Geddes, L. "Nazi Angel of Death not responsible for town of twins." *New Scientist*. Reed Business Information Ltd., 27 Jan. 2009. Web. 25 Apr. 2010. <http://www.newscientist.com/article/dn16492-nazi-angel-of-death-not-responsible-for-town-of-twins.html>

Handwerk, B. "Nazi Twins a Myth: Mengele Not Behind Brazil Boom?" *National Geographic News.* National Geographic Society, 25 Nov. 2009. Web. 29 Apr. 2010. <http://news.nationalgeographic.com/news/2009/11/091125-nazi-twins-brazil-mengele.html>

Matalon-Lagnado, L., Cohn-Dekel, S. *Children of the Flames: Dr. Josef Mengele and the Untold Story of the Twins of Auschwitz.* New York: Morrow, 1991.

Walters, G. *Hunting Evil: The Nazi War Criminals Who Escaped and the Quest to Bring Them to Justice.* New York: Broadway Books, 2009.

23. Morgellons Disease

In this newly described condition, some patients report strange plastic fibers growing from their skin.

Today the skeptical eye focuses on a newly described condition from the medical fringe: Morgellons disease. This is a skin condition in which a painful rash or other open sores appear on various parts of the body, but with a unique characteristic: Found embedded within these sores are colored fibers, apparently made of plastic or other synthetics. Morgellons has created something of a battle line drawn in the sand between sufferers and medical science. Sufferers believe these fibers are being extruded from the body itself, while doctors and psychiatrists generally agree that the fibers come from the environment and are merely being caught in the sores as always happens with scabs.

Skin rashes and sores are one of the physical symptoms of acute stress, and to most doctors who are aware of it, Morgellons appears to be nothing more than this. It's often compared to delusional parasitosis, where the patient believes that the normal itching of a stress-induced rash is caused by unseen parasites living in or on the skin. No parasites are ever found, but some patients tend to react with hostility toward any diagnosis that does not support their preconceived notion. But doctors can only go by the best state of our current knowledge, and are the first to admit that we don't know everything about the human body or about diseases. So to take a truly skeptical perspective, we start by setting aside what we think we know and looking at the evidence, beginning with the history.

Morgellons had a particularly inauspicious beginning. In 2001, a former hospital lab technician turned stay-at-home mom, Mary Leitao, noticed a raw patch under the lip of her two-year-old son Drew. She took him to eight (!) different doctors, dissatisfied with each diagnosis that there was nothing unusual wrong with Drew. She picked fibers from the surface of the scab and examined them under Drew's toy microscope. Her own conclusion was that the fibers were being extruded from Drew's skin, rather than coming from a blanket or stuffed animal or anything else that toddlers bury their faces in. Drawing on the word *morgellons* from an old French reference to black hairs, she created the name Morgellons Disease.

Leitao demanded that the doctors prescribe antibiotics, which they would not do, given the lack of any apparent illness. She became obsessed with finding a doctor who would validate this new disease she'd invented. One doctor at Johns Hopkins wrote to another "I found no evidence of [anything suspicious] in Andrew... Ms. Leitao would benefit from a psychiatric evaluation and support, whether Andrew has Morgellons Disease or not. I hope she will cease to use her son in further exploring this problem."

Another doctor at Johns Hopkins agreed, and even took it a step further, stating that Leitao appeared to be a case of Münchausen's by proxy. Münchausen's Syndrome is where you pretend to be sick because you love getting attention from doctors and hospitals. Münchausen's by proxy is a psychiatric syndrome where you take a child or other family member, and promote them as being sick, to get the same attention. It need not be a conscious deception, Mary Leitao almost certainly does genuinely believe her son is ill; but the psychiatric pathology is the same. She has since gone on to found the Morgellons Research Foundation, which currently lists 14,700 registrants.

An Internet search today reveals that Morgellons has become conflated with chemtrail conspiracy mongering. Some believe that contrails left by airplanes are actually the government spraying toxins to sicken the population with Morgellons. An article on the conspiracy theory web site Rense.com compares two pictures, one claimed to show a fiber from a Morgellons sufferer, and another claiming to show a fiber from chemtrail spraying. It says:

> *Common characteristics of both types of fibers appear to be similar size and chaotic, uncontrolled growth. If these fibers are the result of highly advanced nanotechnology then we have found the disease, and possible who is behind it. But what would be the purpose of forcing this ailment on the population? Torture? To create a new pandemic in order to sell a new drug for a "treatment?"*

Many pro-Morgellons sources claim that the fibers have defied all explanation: They are not human hair, they are not synthetic fiber, and they are not natural plant-based fibers. But I found two significant problems with these assertions. First, they seem to be nothing more *than* assertions, often accompanied by a story that someone looked at them under a microscope and was somehow able to rule out all known fiber compositions. Second, there is little agreement on the characteristics of the fibers, and thus no way such an assertion can be broadly applied. Some sufferers describe hard, solid plastic shards, often in bright colors. Some describe them as thick hairs. The most common photograph on the Internet shows a tangle of fine filaments. Others find curly threads consistent with synthetic fibers from brightly colored blankets, carpet, or sweaters.

So now let's look at the common medical explanation for Morgellons: Acute stress. Acute stress is known to produce all the same symptoms reported by Morgellons sufferers, including painful, itching skin rashes that the patients scratch, producing open sores that capture fibers from clothing and the environment. Stress also results in chronic fatigue, headaches, sleep loss, memory loss, and mood disorders. Do you think Morgellons might be stressful? Here's a description of what it's like to have Morgellons, from Mary Leitao's Morgellons Research Foundation:

> *It is difficult to understand the tremendous suffering caused by this disease. Many patients report feeling abandoned by the medical community, as they experience increasingly bizarre, disfiguring and painful symptoms, while often being unable to receive medical treatment for their condition. A large number of patients become financially devastated and without health insurance because they can no longer work. Most people who suffer from Morgellons disease report feeling frightened and hopeless.*

I've gotten at least a dozen emails from Morgellons sufferers over the past couple of years, and I've also gone to YouTube and watched the reports of dozens more. There is one thing that they nearly all have in common: They almost always say something like "The doctors told me I was crazy, they told me I was imagining it, they told me it was all in my head." In my experience communicating with such people, I've come to doubt that what they were told was actually worded like this. It was probably something like "the causes of what you're going through are usually psychogenic," which the patient misinterprets and exaggerates into a straw man when they retell it. Psychiatrist Alistair Munro, author of *Delusional Disorder*, said "The moment you mention psychiatrists, these patients get extremely angry. They say there's nothing wrong with their brain."

You don't have to be crazy, and you don't have to imagine anything, to experience these symptoms. It's common for people who are perfectly sane and smart to make erroneous self-diagnoses. The wealth of validating, affirming information on the Internet, on sites like Mary Leitao's, exacerbate this problem. Many patients see themselves as "armed for battle" with doctors, with a battery full of information from the Internet. This frequently results in a stalemate. Patients charge that the doctors have closed their minds to the possibility that they suffer from a physical disease, and doctors find that patients have closed their minds to all but their own Internet-supported self diagnosis.

So how do we bridge this gap? Both parties have to open their minds and take steps. The doctors have already done so. In early 2008, the Centers for Disease Control launched a major investigation to learn more about Morgellons, which for now they're calling "unexplained dermopathy". The investigation is currently underway, and there's an email address and a phone number with recorded updates on the latest news. The study is being done through a medical center in Northern California that has an unusually high number of patients reporting the symptoms. For more information, visit CDC.gov/unexplaineddermopathy.

Getting the patients to take steps of their own and open their minds has proven substantially more difficult, for the reasons just discussed. Doctors face the challenge of getting patients to agree to treatment for a diagnosis to which they're often actively hostile. Despite many doctors' best efforts to communicate to their patients that they might benefit from treating the stress that accompanies their symptoms, all too often patients wrongly hear "You're crazy, you're imagining it." It's like a mine field. If I were a doctor, this would be my pitch to a Morgellons sufferer:

> *We don't yet know what causes these filaments and the related symptoms, but the CDC is currently investigating it. We hope to have a proven diagnosis and treatment once they finish their studies. In the meantime, we may be able to make you more comfortable by helping you deal with the stress that this is causing you. Whatever the cause turns out to be, the accompanying stress is making you suffer more than you deserve to.*

Does stress treatment help? The psychiatric literature does contain published accounts of delusional parasitosis being successfully treated with psychotropic drugs and with psychotherapy, but practitioners consistently report the difficulty of getting patients to undergo such treatment. So far there don't seem to be any randomized controlled trials of treatments for Morgellons specifically, but from everything we know about delusional parasitosis and other psychogenic conditions, there's every reason to expect psychotherapy to be equally effective for Morgellons.

The Morgellons Research Foundation does recommend one treatment: the use of long-term antibiotics to treat presumed bacteria. Although bacterial infections are known not to cause strange plastic fibers to grow from your skin, and thus are logically ruled out as the cause, the belief that the antibiotic will help can leverage the placebo effect to reduce or eliminate the stress caused by belief in the disease. In any case where there is no actual disease agent, a placebo of any kind has a good chance of becoming a partial or complete cure. But there are good reasons not to take unnecessary antibiotics, so non-drug therapies would be preferred, if sufferers would be willing to attempt them.

If you really want to help someone who suffers from Morgellons or any other psychogenic condition, you can't only rely on telling them they're wrong and depending on them to take all the steps. But you also don't have to dishonestly pretend their self-diagnosis is true. There is middle ground that you can both reach, if you're both willing to find it.

References & Further Reading

CDC. "Unexplained Dermopathy." *Centers for Disease Control and Prevention.* Centers for Disease Control and Prevention, 29 Jun. 2007. Web. 15 May. 2010.
<http://www.cdc.gov/unexplaineddermopathy/>

Devita-Raeburn, E. "The Morgellons Mystery." *Psychology Today.* Sussex Publishers, LLC, 1 Mar. 2007. Web. 14 May. 2010. <http://www.psychologytoday.com/articles/200702/the-morgellons-mystery>

Harlan, C. "Mom Fights for Answers on What's Wrong with Her Son." *Pittsburgh Post-Gazette.* 23 Jul. 2006, Newspaper.

Koblenzer, C. "The challenge of Morgellons disease." *Journal of the American Academy of Dermatology.* 1 Jan. 2006, Volume 55, Number 5: 920-922.

Koo, J., Lee, C. "Delusions of parasitosis. A dermatologist's guide to diagnosis and treatment." *American Journal of Clinical Dermatology.* 1 Jan. 2001, Volume 2, Number 5: 285-90.

MRF. "Frequently Asked Questions." *Morgellons Research Foundation.* Morgellons Research Foundation, 1 Jan. 2009. Web. 15 May. 2010. <http://www.morgellons.org/faq-home.htm>

Robles, D., Romm, S., Combs, H., Olson, J., Kirby, P. "Delusional disorders in dermatology: a brief review." *Dermatology Online Journal.* 1 Jan. 2008, Volume 14, Number 6.

24. Dinosaurs Among Us

Are some examples of ancient art proof that dinosaurs and humans coexisted?

In this chapter, we point our skeptical eye at the jungles of the Dark Continent, and other remote hideaways throughout the world, where tales tell that living relics from the past still walk among us: the dinosaurs. From Mokele Mbembe, the alleged sauropod of the Congo; to the Ropen, said to be a pterosaur ruling the skies of Papua New Guinea; to the idea that plesiosaurs are the lake monsters of Loch Ness, Ogopogo, and others; the reports come from all over. You'd think these stories would be on the decline. As humans spread out into the farthest reaches of our planet and explore more, you'd expect the stories to fade as nothing is found. However, they're actually on the rise, due to promotional efforts by the relatively new Young Earth Creationism movement intent on proving that dinosaurs lived so recently that they coexisted with humans, and may even survive today.

But all of that aside, this was a chapter I was pretty excited to write, because it's really fun to examine evidence of something so interesting as living dinosaurs. But sadly, I was immediately disappointed. Dig as much as I could, I found that there is no solid evidence for almost any of these animals. There are tremendous volumes of anecdotal stories, nearly all reported by impassioned cryptozoologists, and nearly all based on interviews of native people: secondhand reports of secondhand reports.

But surely these people must be seeing something. Legitimate zoologists who have followed up on the cryptozoologists' claims routinely find that known animals were likely the cause of the stories: birds for the Ropen, and hippos or crocodiles for the Mokele Mbembe. Since the personal anecdote route has failed to produce hard data, cryptozoologists and Young Earthers have turned to ancient artwork in an effort to form a parallel line of evidence. Chief among these accounts is a stone carving buried in the jungles of Cambodia, the Buddhist temple of Ta Prohm.

The Ta Prohm Stegosaurus

Ta Prohm is often featured in popular culture. It's best known for its jungle trees growing among the moss-green stone ruins, most famously for the great roots flowing over it that look like they were poured into place, and its giant stone faces of Buddha. Virtually the entire temple is carved with Buddhist images or decorations. Of particular interest is one column tucked away in a corner, graced with a winding serpent that encircles a number of animals. Some are recognizable as actual animals, others are chimera or mythical creatures such as garudas or nagas. But one stands out in particular, because at first glance, you might think it looks like a Stegosaurus. It's a stout four legged animal, its big head hanging low, with a tail about like that of a dog. Most significantly, along its back is a row of pointed plates.

The Ta Prohm Stegosaurus has made waves throughout the cryptozoology world, appealing not only to those who believe that relic dinosaurs still exist in parts of the world, but even to Young Earth Creationists desperate for evidence that humans and dinosaurs coexisted. However, upon any reasonable inspection, the Ta Prohm creature fails to serve as good evidence of either of these hypotheses. There are at least three dramatic differences between it and a Stegosaurus. First and most significantly, Stegosaurus had a tiny head, such that from a side view, it's hard to tell which end is its head and which is its tail. Both were long, graceful, and tapered out to a point. The Ta Prohm creature, conversely, has a massive head, perhaps a quarter the size of its entire body, like that of a hippo, and no neck to speak of. Second, the Ta Prohm creature is completely missing Stegosaurus' most identifiable feature: the thagomizer, the collection of four spikes at the tip of its tail. Finally, the distinctive plates rising from the spine are all wrong. Stegosaurus had 17 plates of greatly varying size, tiny at the head and tail, rising to very large at the top of the back. Ta Prohm has only six or seven, all of equal size. If the Ta Prohm carving did indeed use a living Stegosaurus as its model, then its quality is grossly out of step with that of all the other animals carved at Ta Prohm, which are quite accurate and beautifully done.

Javan rhinoceros

Of course, we can't know what was in the mind of the artist. But we can get an idea from looking at all the carvings in context. In all of the backgrounds, foliage is depicted. The Stegosaurus would be the only animal shown without accompanying foliage, unless we make a different interpretation of the image. If we interpret the "back plates" as background foliage, to bring the image in line with all the others, we're left with a common, fairly generic quadruped. I think it looks a lot like a single-horned Javan Rhinoceros that lived in the region at the time of Ta Prohm. Other identifications have been a wild boar, or even a chameleon. None of these are perfect matches, but all are much closer than Stegosaurus, and all are real animals that would have been well known to the Ta Prohm artists. And, of course, since Ta Prohm depicts many mythical beasts, there isn't even a need to identify the creature as a real animal. That Stegosaurus must have lived in Cambodia only 800 years ago drops to among the least likely of many possible explanations for the carving.

But even given its weaknesses, the Ta Prohm creature is head and shoulders above the rest of the evidence that's in the form of ancient art. At the bottom end of this spectrum is a formation from Bernifal Cave, one of the many caves in France filled with Cro Magnon pictographs. The paintings in Bernifal all show real animals, but some Young Earthers point to part of the rock surface that they believe has been carved to show a generic dinosaur butting heads with a mammoth. This is one of those cases of pareidolia, like the Face on Mars. It's fair to say that the contours on the rock do vaguely look like the head and jaw of a dragon-like creature, but what they call a mammoth is just a blob. There are no legitimate petroglyphs in Bernifal. (A pictograph is painted on the surface of rock; a petroglyph is made by chipping into the surface.) This alleged battle to the death has the same surface texture as the rest of the cave, the same general contouring, and has never been included in any legitimate archaeological survey of the cave's artwork.

But there is real artwork, that human artists actually did make, that's better. Virtually every culture throughout history has produced art, much of it very high quality, that depicts dragons or other beasts that look something like some prehistoric species. I could speak for hours simply listing the excellent examples from China, Egypt, Mesopotamia, Rome, Mesoamerica, even North America, where you could point directly to a known dinosaur species and make a match. But you have to understand how illogical it is to consider any of this art to be evidence that the depicted creature was actually known to the artist to be a living or real animal. Art, by definition, is a representation of the artist's imagination or impression. There is an even larger number of artworks from all of these cultures that even Young Earthers and cryptozoologists would readily admit were not intended to be photorealistic representations of actual living beings. Some fantastic creatures in art happened to resemble real species; many more did not. Ancient artists did not employ a flagging system to unambiguously tell us which of their art represented mythical beings and which were intended as historical records of living animals.

I'll give two specific examples that I think would be among the most convincing: A pair of long-necked dinosaurs engraved in brass on the 1496 tomb of the Bishop of Carlisle in the UK, and another stegosaurus (of much more accurate proportions than the one at Ta Prohm) on a shard of ancient Greek pottery found in modern day Turkey. Now, it's possible to debate the details of these works all day long. Neither of them quite match what we now believe these animals looked like, including some very significant anatomical differences. But this line of reasoning is never going to get you anywhere; you can argue yourself into circles all day long and never change the mind of someone who believes that if an ancient piece of artwork superficially matches a known dinosaur, it's therefore evidence.

This all comes down to the value of anecdotal evidence. A personal account, whether it's a verbal story, a sketch, a written report, or a stone carving, cannot be tested. No matter how authoritative or reliable we consider a witness to be, his account, by itself, cannot be validated scientifically. The line of reasoning that *Someone told a story, therefore it must be true* is precarious indeed. The reverse is just as invalid: *Someone told a story, therefore it must be false.* There are so many other possibilities: Fiction, legend, metaphor. And significantly, mistaken interpretation is just as possible on the listener's end as it is on the teller's end.

If we do find a Ropen or a Mokele Mbembe one day, it seems likely that their numbers will be pretty small. Maybe relic dinosaurs were around in larger numbers when some of these ancient artists were active, but all the testable evidence we have for dinosaurs places them tens of millions of years before the first protohuman stood up. We have only anecdotes that suggest otherwise, anecdotes that fail to be backed up by the testable evidence that we would expect to exist were these creatures real.

Do dinosaurs survive in some remote corner of the world? I certainly hope so, and I think most people would love for it to be true; but I'm not putting my money on it. I think a dinosaur would be pretty hard to miss. Don't let your emotions govern your science. No matter how much you want something to be true, always consider the quality of the evidence. If it's anecdotal and unsupported by corroborating testable evidence, you have very good reason to be skeptical.

References & Further Reading

Bahn, P. *Cave Art: A Guide to the Decorated Ice Age Caves of Europe.* London: Frances Lincoln Ltd., 2007. 66-67.

Carpenter, K. *The Armored Dinosaurs.* Bloomington: Indiana University Press, 2001. 76-141.

Hall, A. *Monsters and Mythic Beasts.* London: Aldus Books, 1975. 84-99.

Isaacs, D. *Dragons or Dinosaurs.* Alachua: Bridge-Logos, 2010. 90-96.

Novella, S. "Ancient Cambodian Stegosaurus?" *Neurologica Blog.* New England Skeptical Society, 19 Feb. 2008. Web. 20 May. 2010. <http://www.theness.com/neurologicablog/?p=196>

Woetzel, D. "Ancient Dinosaur Depictions." *Genesis Park.* Dave Woetzel, 9 Mar. 2002. Web. 20 May. 2010. <http://www.genesispark.org/genpark/ancient/ancient.htm>

25. The Westall '66 UFO

200 students watched a strange craft fly near their school in Australia in 1966. What did they see?

Melbourne, Australia, 1966. A sunny, breezy day in autumn, April 6 to be exact. Field sports were underway for a morning class at Westall High School. A few students saw it first, and then a few more. They described it as a disk, gray or silver, about the size of two family cars, and about four football fields away. It hovered silently, and then descended out of view behind a row of pine trees to the south of the school. A few minutes later it emerged, only now it was being pursued by a squadron of five light aircraft, and now its movement was faster. The object, now described as a small, bright streak of light, darted about with the aircraft playing a game of cat and mouse. After 20 minutes the strange object and the airplanes pursuing it went out of view. As soon as they had the chance, some students scrambled toward the trees and found the grass flattened where the object had undoubtedly landed while it was out of view. Back at the school, students and staff were instructed not to talk about what they'd seen. Intimidated by the sight of military personnel, the students have allegedly remained silent ever since.

But as we almost always see with urban legends, the more time passes, the larger the story grows. 44 years after the event, retellings have expanded significantly compared to what was documented at the time, and this should always give us cause to approach modern revisionings with skepticism. New evidence coming to light is one thing, but with the Westall event, all we see are new anecdotes. A few adults who were students at the time, and UFO proponents who have interviewed former students, are now reporting greatly expanded versions of what happened. Does that make them wrong? Of course not. But if we want to determine the most likely account of what really happened, we go to the original sources. We go to the original documentation of what the witnesses reported 44 years ago, and we take the contradictory revisionings with a large grain of salt.

Let's start by having a look at the area. Westall High School, now called Westall Secondary College, is in a suburb of Melbourne, Australia called Clayton South. The land all about Westall is quite low and flat, and the coast is just about 10 kilometers southwest. Just to the south of the school was a natural open space called the Grange Reserve, mostly trees and scrub. Most of the Grange is still open space and remains today, including the row of trees visible from the school behind which the object was seen to descend. Beyond the Grange, about 4.5 kilometers, is Moorabbin Airport, which was then and still is a small but very busy general aviation airport. By number of takeoffs, it is in fact the third busiest airport in the southern hemisphere.

Its proximity to the school has contributed to both UFO researchers and skeptics suggesting the sighting may have been of an experimental military craft. However, I don't find this explanation very convincing at all. Australia didn't really have much of an aircraft industry in 1966. They'd been quite busy during World War II, but by the 1960's it was scaled way back, and most of Australia's aircraft industry was providing service and support to existing planes. There were a few exceptions. The Commonwealth Aircraft Corporation was active at the time, but they did no design work, they merely constructed models in use by the Australians that had been designed overseas. At the time of Westall, they were busy producing the Dassault Mirage III fighter and were just ramping up for production on the Maachi MB-326 trainer. De Havilland Australia had developed two prototype jet trainers, the P17 and the F2, but both of these projects were cancelled in 1965 when the company shifted its focus to production of parts for Commonwealth's MB-326. The only other aircraft manufacturers in Australia were tiny and built only small civilian or agricultural planes. In 1966, none of these companies had anything like the skunkworks of American aircraft companies that we usually think of when we talk about strange experimental craft.

Of course, other countries did, not only the Americans, the Canadians, and the British, but also France, the Soviet Union, and others. Certainly any of them might have chosen to test advanced designs in Australia. The problem with this hypothesis is that all such designs have long been declassified and are now well known, and none of them would be a serviceable match for the Westall reports. Nobody had anything that hovered silently, darted about playing cat and mouse, or flattened grass when it landed. The closest thing I can find would be the flying saucer shaped Avrocar, which was desperately unsuccessful and had been cancelled five years before Westall. If we introduced the suggestion that maybe an improved version lived on in secret and was tested in the middle of Melbourne in broad daylight, we'd be on very thin ice. There's no evidence for that, and it would remain unknown to every aviation historian and author.

There was one strange craft launched that day, however: A weather balloon, reported the next day in the newspaper *The Age* as a possible explanation for the event. It was launched from Laverton two and a half hours before the sighting, 32 kilometers west-northwest of Westall. *The Age* reported that the wind was blowing from the west, and if it continued southeast near Clayton South, the balloon could likely have disappeared from view behind the row of trees, very close to 11:00am when the sighting happened. Despite being dismissed by UFO promoters, I find this balloon event to be a very plausible candidate for the first half of the sighting, when a silently hovering disc descended behind the trees.

The second half of the event had a much different character. Andrew Greenwood was a science teacher at Westall High

School, and is the only staff member known to have reported seeing the object at the time. He gave a detailed account to the newspapers. Greenwood first saw the object when it rose into view from beyond the pine trees at the Grange. He described it as a silvery streak "like a thin beam of light, about half the length of a light aircraft." At first it appeared with only a single aircraft, but was eventually joined by five. He described a "cat and mouse" game that the aircraft played with the object, a game which lasted a full 20 minutes. The object moved side to side and its size appeared to fluctuate slightly. By the end of morning recess, Greenwood said he turned away, and when he looked again the object and the five airplanes had gone.

One man, who identified himself only as a former RAAF navigator, wrote a letter to the editor in the April 28 *Dandenong Journal*, in which he said that Greenwood's report was a "reasonably accurate" description of a nylon target drogue, like a wind sock, towed by one plane for the others to chase, and known to be in use by the local RAAF at the time. A "cat and mouse" game would be a fairly apt description of what happens when pilots undergoing training try to follow the drogue. For an explanation of why no pilots reported anything strange, he offered "Why should they? They were probably carrying out a normal...exercise and wouldn't dream that anyone could take a drogue for a 'flying saucer.'"

Although this sounds to me like a spot-on explanation for the second half of the sighting as reported in 1966, there's no evidence that anyone was conducting any drogue exercises there at the time. There's no evidence that they didn't, but we can't do any better than list this as one possibility.

Talk of military records leads us to the alleged secrecy that was imposed following the event. Did military personnel show up and silence everyone to cover up the event? It does not seem likely, since newspapers widely reported the story, and everyone who's ever been interviewed about it has spoken quite freely; there is nothing about this story consistent with any kind of coverup having taken place. On April 14, the *Dandenong Jour-*

nal reported that the school would not permit any further interviews with students, and that students and staff at the school had been asked not to talk to reporters. Was this evidence of a government conspiracy? The principal, Frank Samblebe, gave a simpler explanation to the *Dandenong Journal*, published on May 5, 1966. He said "the flood of callers and phone calls from the Air Force down to the Flying Saucer Association interrupted the children's studies." Given this real-world concern of the school, it does seem reasonable that he would have asked the press to leave the students alone, and done what he could to enforce it at the front desk. From every single 1966 account I've read, this is the full extent of what's now being described as a "coverup" or a "conspiracy".

The Air Force personnel Samblebe referred to were probably four RAAF investigators who showed up on April 9, three days after the event, to look at what was said to be the landing site. A number of enthusiasts from various UFO groups accompanied them, but apparently nothing interesting was found, because nothing was documented from this visit. The newspaper interviewed one student who said she thought the tall grass looked merely like the wind had flattened it. Indeed the flattened circles were apparently so vague that witnesses couldn't agree whether it was one or three. Some reports say the RAAF men burned the area to hide the evidence, but according to the farmer who owned the land, he burned it himself to stop people from trampling onto his property. Today's expanded accounts often include much deeper military involvement, such as police and military "swarming" around the school and cordoning off the "landing site", and the circles appearing to be "scorched", but I didn't find a single record from 1966 to substantiate this.

Something else that's grown over the years has been the number of witnesses at the school. The *Dandenong Journal* reported at that time that only one teacher and "several" students saw the object, but by now that number has grown to 200. This number is probably artificially inflated by what behavioral psychologists call the bandwagon effect. When all of your friends say they saw something, you tend to say you saw it too, whether you really did or not, simply because you don't want to be left out or be considered inferior. Most of the witnesses were high school students, and that's an age when we tend to be highly conscious of our image and social conformity. Nobody wants to be the one who couldn't see the object. Some of the students most likely did see something, but it's fairly certain that at least some of the witnesses simply went along with the crowd in accordance with the bandwagon effect.

This introduces a serious problem. The descriptions of what was actually seen have now become diluted with made-up descriptions by an unknown number of students who didn't see anything, and there's no way to know which is which. Police investigators are keenly aware of this potential complication, and often have to account for it. In practice, this usually means finding commonalities among the accounts of witnesses who were in the best positions, and discounting aberrant reports, usually from those who were not in as good a position, and especially from those who come forward later after the general facts have become publicly known.

Employing this strategy, we can bring what probably happened at Westall into better focus. The commonalities of the reports from the witnesses in the best position all state that the craft was a great distance away, beyond the trees. This is probably what happened. Aberrant reports from a few students, such as those who report that it flew over the school, or that it landed in the schoolyard, or the one girl who said she touched the craft as it lifted off, are much less reliable and probably safely discarded. Such reports are sensational and most likely to make headlines, but to the investigator who knows his business, there is good reason to dismiss them. We can say with pretty good certainty that whatever the object was, it was too far away to easily judge its actual size. The best indicator we have of scale is of its second appearance, after it rose from behind the trees, when Andrew Greenwood reported it played cat and mouse with the light aircraft: Small and thin, and shorter than a light aircraft.

So what can we conclude about the Westall UFO? Not very much. The weather balloon is a likely explanation for the first half of the event, and the drogue is at least one very reasonable possibility for the second half. There's good reason to doubt that many of the story elements, like the military conspiracy and the craft having landed, ever happened at all. The story certainly has no holes in it that can only be filled with extraterrestrial aliens, and indeed no credible reason to suggest anything unusual. "I don't know" does not mean "I do know, and it was a spaceship", so for now, the Westall '66 UFO remains one of many question marks in the books, just not a very bold or especially intriguing one.

REFERENCES & FURTHER READING

Anderson, P. *Mustangs of the RAAF and RNZAF.* Sydney: A. H. & A. W. Reed, 1975.

Anonymous. "Mystery Solution?" *The Dandenong Journal.* 28 Apr. 1966, Letters to the Editor: 21.

Beaufort. "The Story Of The Commonwealth Aircraft Corporation." *Beaufort Restoration.* The Beaufort Restoration Group, 2 Sep. 2007. Web. 29 May. 2010. <http://www.beaufortrestoration.com.au/Pages/ProductionChild/Manufacturers/CAC.html>

Bell, D. *The Smithsonian National Air and Space Museum Directory of Airplanes, Their Designers and Manufacturers.* Washington, D.C.: Stackpole Books, 2002.

Boeing. "Boeing Australia." *Boeing.* The Boeing Company, 7 Feb. 2006. Web. 27 May. 2010. <http://www.boeing.com.au/ViewContent.do?id=47382&aContent=History>

Damian. "Around Clayton." *The Dandenong Journal.* 5 May 1966, Editorial.

Editors. "Who Were 5 Pilots?" *The Dandenong Journal.* 21 Apr. 1966, Volume 105, Number 30: 1-2.

Editors. "Object Perhaps Balloon." *The Age*. 7 Apr. 1966, Newspaper: 6.

26. The Lost Ship of the Desert

The facts behind tall tales from the American southwest of ships found in the middle of the desert.

It was 1933, and Myrtle Botts was traveling with her husband, enjoying the famous annual wildflower bloom in what's now Anza-Borrego State Park, in the Colorado Desert of inland southern California. They met an old prospector who swore he'd seen the remains of a Viking longship protruding from the side of an arroyo, well enough preserved that the distinctive round shields were still mounted along its sides. He wrote directions for Myrtle on how to find it. A paper, purporting to be those original directions, is preserved at the Julian Pioneer Museum in Julian, California.

Following the directions, Myrtle went and found the ship. She returned to fetch her husband along, but before they could get there, an earthquake brought down the wall of the arroyo, access was blocked, and the ship has not been seen since. Erosion in the arroyo has since washed away whatever the directions may once have led to.

How could a Viking ship end up in the deserts of the American southwest? Coastal mountain ranges eliminate the possibility that it came from the Pacific Ocean. There is, however, one route into the desert, via the Colorado River, which feeds into the Gulf of California, also known as the Sea of Cortez, between Baja California and Mexico. Historically, this was largely a navigable river for shallow-draft boats, and it would have been possible to sail from the ocean up through what's now Yuma, AZ right on the border with California. Today, however, that river no longer exists, at least not in a very meaningful way. By the time the river gets to Yuma, it's gone through a number of dams and been channeled into numerous agricultural canals. Feeding the dry southwest leaves little water in the Colorado's bed, and in most places between Yuma and the Gulf, the bed is dry.

But before agriculture, it was a mighty channel, so big that it occasionally overflowed its banks and flooded the desert. Most notably, this happened in 1905 when the river burst its banks in Mexico, and flowed northwards in two separate channels for more than two years. The entire Colorado River poured into the Salton Sink in California, filling it to become what's now called the Salton Sea. These two rivers still run today, draining the town of Mexicali into the Salton Sea.

Myrtle Botts' prospector told her the Viking ship was in the badlands west of Mexicali. The land about there is either flat desert or hard rock mountain ranges, except for the Carrizo Badlands. If the Viking ship was in the Carrizo Badlands, it would have had to climb at least 200 meters in elevation. The arroyos, or canyons, in the Carrizo Badlands don't flow toward Mexicali, however. On the rare events when they contain water, they flow north as well, downhill into the Salton Sink.

So it appears that the Salton Sea is the key to any ancient ship traffic that may have left a shipwreck in the southwest. But even this is problematic. We do have good paleohydrology on the Salton Sink. Many times, since about the year 700, the Salton Sink has filled when the Colorado River overflowed into it. The largest was in about 1500, when Lake Cahuilla (as it was then called) was 26 times the size of today's Salton Sea, and its shoreline is still visible on the surrounding hillsides. These overflows happened when the Colorado River silted up its normal path to the Gulf, and naturally diverted itself. At its highest, Lake Cahuilla actually covered what's now Indio, California to Yuma, Arizona, and was contiguous with the Gulf of California. During these events, shallow-draft ships could indeed have traveled inland as far as the outskirts of the Carrizo Badlands.

But historically, this was quite rare, and was unknown to any Europeans who left records. The first known Europeans to the region were led by Melchior Diaz in 1540. He traveled upriver from the Gulf, and then sent expeditions overland to Lake Cahuilla, so we know there was no water connection. By 1604, New Mexico had a Spanish governor, Don Juan de Ornate, and we know from his explorations of the Colorado River that it was not connected to Lake Cahuilla at that time either. 100 years later, the Colorado silted up again, blocking access from the sea, and refilling Lake Cahuilla. But by 1774, the diversion had corrected itself and Lake Cahuilla was dry again, as reported by explorer Juan Bautista de Anza.

This casts doubt on stories of Spanish galleons found high and dry in the desert, but it does leave the door open to other earlier explorers who did not necessarily leave records. Candidates we have for this are the Chinese and the Vikings. In the 1400's, Chinese explorer Zheng He traveled west from China to Africa, but no reliable accounts show him ever venturing east across the Pacific toward America. One map turned up several years ago, purporting to show that Zheng He had visited and mapped America, but it's been shown to be a fake.

Vikings were well known for their wide-ranging exploits launched from Scandinavia, even going so far as northern Africa. They also made several attempts to colonize North America along its eastern coast, but they never ventured anywhere near the Pacific Ocean, so far as we know. Stories have persisted for almost two centuries of blond Eskimos, generally thought to be the descendants of Inuit who intermarried with Vikings along Canada's northern coast. If Vikings had made it that far, we might hypothesize that they could have continued south through the Bering Strait and eventually reached Mexico. But we strike at least two stumbling blocks. First, there is no archaeological evidence of Vikings anywhere along the American west coast, which would necessarily exist; and second, DNA testing of the European-looking Cambridge Bay Inuit with Norse descendants from Iceland proved no genetic link. These combine to make the story of a Viking longship in the Colorado Desert highly unlikely.

But of course, it could be some other kind of boat that Myrtle Botts' prospector mistook for a Viking longship. There are other stories of ships in the desert. One holds that a Spanish galleon was washed out of the river during a flood in 1862 where it became beached some 60 kilometers north of Yuma, but this is probably just a tall tale. It's unlikely that anything as large as a galleon was ever able to navigate up the Colorado. If it was a smaller boat, it's a perfectly reasonable story, as there are floodplains in that part of the river where a boat could quite easily get stuck at high water. But it would not be a terribly remarkable event, and there isn't anywhere around there where a boat could be washed more than a couple hundred meters from the channel.

In the V formed by the south-flowing Colorado and the north-flowing channels into which the Colorado has spilled into the Salton Sink is some high ground with sand dunes called the Sand Hills, and somewhere in these dunes has been reported a Spanish pearl ship. The story goes that a tidal bore pushed the small ship over the shallow delta in 1615, when Lake Cahuilla was so high that the whole area was underwater and the lake was contiguous with the Gulf. It's reasonably plausible that a small ship could have made it there and become stranded, but the Sand Hills are too high and too far from Lake Cahuilla's maximum level. Once the water receded the ship would have been well out in the open and easily found.

That's the case with the plausible version of all these stories. If boats did indeed get stuck in what's now desert, they would have been on the surface, out on the flats, and quite obvious. We can say with pretty good certainty that no ship within the past few thousand years could have become buried in such a way that a more recent canyon could have exposed them. We have a thorough understanding of the geological history of the Carrizo Badlands. The top layers, where any embedded boats might be found, consist of layers of alluvium, from several different formations, laid down during the Pliocene and Miocene epochs. The newest deposits are at least 2.5 million years old, and they get much older from there. So there's really no chance that any kind of boat or other human construction might be found protruding from the wall of a canyon in this region.

We should also keep in mind that the southwest is traveled today much heavier than it was in the 1800's, plus we have the benefit of two centuries of exploration since then. In all that time, we've never had anything better than tall tales; nobody's yet presented evidence of a Viking ship, Spanish galleon, or any other unexpected ancient boats turning up in the southwest where they shouldn't be. If they do, it's likely going to be from a small boat where we would expect such a craft to have been stranded or abandoned in a place where high water is known to have been. Paleohydrology shows us that may indeed be a great distance from any existing shoreline. What will get me excited will be a Viking ship, anything galleon sized, or anything where it shouldn't be. That will be a great day to in search of the lost ship of the desert.

References & Further Reading

American Geographical Society: Approximate Status of 1933. "Map of the Colorado Delta Region." *SDSU Center for Inland Waters.* San Diego State University, 1 Jan. 1936. Web. 4 Jun. 2010. <http://www.sci.sdsu.edu/salton/CoRDeltaFull.JPEG>

Bishop, G., Marinacci, M., Oesterle, J., Moran, M. *Weird California: Your Travel Guide to California's Local Legends and Best Kept Secrets.* New York: Sterling Publishing Company, 2009. 64.

CBC. "DNA tests debunk blond Inuit legend." *CBC News.* CBC/Radio-Canada, 28 Oct. 2003. Web. 3 Jun. 2010. <http://www.cbc.ca/news/story/2003/10/28/inuit_blond031028.html>

Lovgren, S. "'Chinese Columbus' Map Likely Fake, Experts Say." *National Geographic News.* National Geographic Society, 23 Jan. 2006. Web. 3 Jun. 2010. <http://news.nationalgeographic.com/news/2006/01/0123_060123_chinese_map.html>

Reheis, M., Hershler, R., Miller, D. *Late Cenozoic Drainage History of the Southwestern Great Basin and Lower Colorado River Region: Geologic and Biotic Perspectives.* Boulder: The Geological Society of America, 2008. 368.

SSA. "Historical Chronology." *Salton Sea Restoration: 13 Years of Progress.* Salton Sea Authority, 1 Jan. 2000. Web. 2 Jun. 2010. <http://www.saltonsea.ca.gov/histchron.htm>

Weight, H. *Lost Ship of the Desert: A Legend of the Southwest.* Twentynine Palms: The Calico Press, 1959.

27. THE NORTH AMERICAN UNION

Are the United States, Canada, and Mexico planning to merge into a single huge police state?

In this chapter, we're going to put on our cheap suits, stick earpieces in, and join the legions of multinational Secret Service agents flowing out among the populace of Canada, the United States, and Mexico; as the borders disappear and we round up a unified population into forced socialism under martial law in our gigantic new pancontinental police state. Some believers say this takeover is actually already underway; others reckon the plans are still being laid, but few believers doubt that it's in the works. The ultimate goal, according to the rumors, is for the few elite in the new government to enjoy unprecedented power, control, and profit over a new supermassive megastate, at the expense of half a billion workers forced into socialized labor. This new police megastate will be called the North American Union, or NAU.

One common theme in the conspiracy rumor is that anytime something bad happens for real, it's generally viewed by the believers as a deliberate attack by the American government upon its own people, as part of this active, ongoing process. 9/11 is the most obvious example; the conspiracy community believes nearly absolutely that 9/11 was perpetrated by the government, in part as an excuse to increase domestic control in preparation for the NAU takeover. When Hurricane Katrina killed over 1,800 people in 2005, some in the conspiracy community took it as an actual practice exercise by FEMA to round up thousands of young men and execute them in the swamp. Even the 2010 explosion of the oil rig *Deepwater Horizon* and subsequent oil spill was described by some as a deliberate attack, with the dual goals of damaging the local economies and reaping huge profits for the government insiders through stock manipulation.

One piece of evidence the believers claim proves that the NAU takeover is underway is a new monetary unit to replace the dollar, called the amero (obviously patterned after the euro used in the European Union). It should be stressed at this point that all known examples of actual amero bills or coins have been proven to be hoaxes; there is no such thing (so far as we know) as actual amero currency. But conspiracy theorists can be well excused for suspecting that an amero is in the works. After all, the euro certainly became a reality in Europe; therefore the amero might well happen here, so they reason.

However, comparisons between the amero and the euro, or the North American Union and the European Union, do not hold much water. The euro was first planned as a solution to a number of problems unique to Europe, where there were many small countries that necessarily had to do large amounts of business among one another; but with all their own separately fluctuating currencies, there were all sorts of problems. Foreign investment was unnecessarily complicated and troublesome, exchange was inefficient and costly, interest rates were unpredictable, and various inflation rates made every transaction a shot in the dark. When the euro was introduced, participating countries had to meet certain requirements of stability. Since its introduction, studies have found that it was tremendously successful in addressing the problems. As of 2006, the European Central Bank estimated that the euro increased trade among member nations by 5-10%, and later estimates show that this trend has only continued to improve.

North American countries, by contrast, do not have anywhere near the currency-related problems that Europe faced before the euro. We simply don't have an unmanageable number of international transactions suffering from fluctuating exchange rates and expensive conversions. Mexico suffered these problems historically, but when the North American Free

Trade Agreement (NAFTA) was signed in 1994, it had a rocky start, but followed by a hugely stabilizing effect on Mexico's economy and ever since, their problems have been largely mitigated. Arguably, NAFTA has addressed many of the issues for North Americans that Europeans solved with the euro. We just had those problems on a far smaller scale, and so our fix was correspondingly less drastic.

But that doesn't mean nobody's ever proposed an amero. There's been talk of it for a long time, mainly from a small number of Canadian economists. Quebec is one faction in North America that would actually benefit from an amero, and Mexico is probably the other. But since the United States and the majority of Canada would not, the amero is unlikely to ever proceed beyond the ruminations of these few authors.

In the real world, the introduction of an amero would probably have real benefits for a few, but none for the majority. Historically, the amero's proponents have been Quebec and Mexico. Quebec's perspective is an interesting one. They have a certain degree of French-Canadian nationalism, and being part of Canada and tied to its currency rubs this nationalism the wrong way. If they shared an amero with everyone else in North America, it would make them less dependent on Canada economically and freer to trade directly with the United States. Former Mexican President Vicente Fox expressed his desire for an amero openly, on multiple occasions, as a natural followup to NAFTA. Such economic unions often confer more benefit upon those at the bottom of the food chain that those at the top. Mexico would benefit from increased stability, while Canada and the United States would lose control of their own inflation and interest rates.

Canadian professor of economics Herb Grubel wrote a 1999 paper for the Canadian think tank The Fraser Institute called *The Case for the Amero*, but in it he admitted that his arguments were probably less important to the governments of Canada and the United States than the need to maintain control over their own monetary independence. The other most significant proponent has been Dr. Robert Pastor, professor of political science and former national security advisor under President Jimmy Carter. In his 2001 book *Toward a North American Community*, he pointed out the benefits of an amero for Latin America, but failed to convince very many people that it had any benefits for the United States. He did admit in the book that an amero was unlikely to happen, and has said that he absolutely does not support a North American Union.

So with academic and economic expertise in agreement that neither a North American Union nor even an amero make much sense, what support remains? Well, unfortunately, it's really just of the conspiracy theory variety, drawing its evidence from misinterpretations and exaggerations of actual events.

One such actual event feeding the conspiracy theory is a group formed by businessmen and academics from Canada, the United States, and Mexico called the Independent Task Force on North America. It's sponsored by nonprofit think tanks from all three nations. Created in 2004 in a post-9/11, post-NAFTA world, it advocates closer economic and security cooperation among the three nations, generally all good ideas. They do not advocate either an amero or a North American Union, all of their reports are freely available, and there's no secrecy attached to anything they do. Nevertheless, some conspiracy theorists consider their existence to be proof that the North American Union is already happening.

A similar, but more official, group was called the Security and Prosperity Partnership of North America, formed in 2005 by Vicente Fox, George W. Bush, and Paul Martin, who met for dialog for essentially these same purposes, and was active through 2009. It included no treaties or agreements. The US web site for the SPP says in its "Myth vs. Fact" section:

> ...*The SPP [seeks] to make the United States, Canada and Mexico open to legitimate trade and closed to terrorism and crime. It does not change our courts or legislative processes and respects the sovereignty of the United States, Mexico, and Canada. The SPP in no way, shape or form considers*

the creation of a European Union-like structure or a common currency.

But nevertheless, you can say this up and down and standing on your head; the diehard conspiracy theorists dismiss such a statement as just another part of the coverup. Notably, in June of 2006, CNN anchor Lou Dobbs described the SPP, on the air, as an agreement (which it wasn't) to actually form the North American Union without the consent of Congress (which it didn't).

> *...The Bush administration is pushing ahead with a plan to create a North American Union with Canada and Mexico. You haven't heard about that? Well, that's because Congress hasn't been consulted, nor the American people.*

Dobbs is not the only one. Believers all across the Internet say "Hey, it happened in Europe; it can happen here in North America." The European Union is indeed a reality, so by that example, it's plausible that a North American Union could happen as well, right? The European Union is actually *not* a real-world precedent for what the North American Union is believed to be. The EU is primarily an economic union. All the member nations in the EU are still sovereign nations, holding their own independent elections and issuing their own passports, and no European citizens are being forced into labor camps or executed by the millions. Conversely, the claims about the North American Union have the United States, Canada, and Mexico merged into a single police state characterized by brutality and forced socialism.

The former Soviet Union would be a closer precedent, but still not a very good one. It *was* a police state characterized by brutality and forced socialism, but there are two very important points to heed. First, the Soviet Union was the result of a popular uprising by the people, the Russian Revolution; it was not a secret takeover by hidden Illuminati intent on deceiving the masses. The Bolsheviks were a majority party, and there was nothing secret about them. Second, the Soviet Union didn't last, and remains a dramatic example of why such a union is a bad idea for everyone; not just for the people, but for everyone hoping to benefit from it.

Like all conspiracy theories that claim to predict future events, the North American Union requires reliance on supposition and irrational dismissal of evidence. Anyone who thinks the United States is likely to give up its sovereignty has, shall I say politely, "lost a few tiles on re-entry." Ask healthy questions and maintain a healthy skepticism; but if you catch yourself departing a little too far afield, take it as a red flag and point your skeptical eye at yourself as well.

References & Further Reading

Baldwin, R. "The Euro's Trade Effects." *European Central Bank Working Paper Series.* 1 Mar. 2006, Number 594: 48.

Chintrakarn, P. "Estimating the Euro Effects on Trade with Propensity Score Matching." *Review of International Economics.* 1 Feb. 2008, Volume 16, Issue 1: 186-198.

Dobbs, L. "Lou Dobbs Tonight (Transcript)." *Cable News Network.* Time Warner, 21 Jun. 2006. Web. 8 Jun. 2010. <http://transcripts.cnn.com/TRANSCRIPTS/0606/21/ldt.01.html>

Grubel, H. "The Case for the Amero." *Fraser Institute Critical Issues Bulletin.* 1 Jan. 1999, 1999 Edition.

Pastor, R. *Toward a North American Community: Lessons from the Old World for the New.* Washington, DC: Institute for International Economics, 2001.

SPP. "SPP Myths vs Facts." *Security and Prosperity Partnership Of North America.* US Department of Commerce, 21 Aug. 2006. Web. 9 Jun. 2010. <http://www.spp.gov/myths_vs_facts.asp>

28. Attack on Pearl Harbor

Did the American government have advance knowledge of the Pearl Harbor attack, and allow it to happen?

Every schoolchild knows the story of December 7, 1941, "A date which will live in infamy". Japanese aircraft carriers crept to within striking distance of Hawaii and launched a morning sneak attack that struck at 7:55am. Two waves of 354 Japanese bombers, dive-bombers, torpedo bombers, and fighters decimated an unprepared U.S. Pacific Fleet. They sank four battleships and two destroyers and heavily damaged eleven other ships, destroyed or damaged 343 aircraft, killed 2,459 servicemen and civilians, and injured 1,282 others. Less than 24 hours after the first bomb fell, the United States declared war on Japan. One question has plagued the conspiracy minded ever since: Was the United States truly caught by surprise, or did the government have advance knowledge of the attack and allow it to happen, as an excuse to declare war?

We should begin by establishing that the overwhelming majority of historians are not moved by this theory. It is promoted really only by a few authors and anti-government activists. However, that doesn't make it wrong. Most Americans have heard the theory suggested, usually in the context of it being an open question. It's not. The jury is not "out" on this one, despite a tiny minority of amateur historians making a majority of the noise. But as we always do on Skeptoid, we'll give the fringe their day and look at their evidence.

Perhaps the most popularly known clue is that the United States' three aircraft carriers were safely out of harm's way. They were out on maneuvers, and were not in port in Pearl Harbor with the rest of the Pacific Fleet. If the American commanders wanted the attack to happen, they would probably still choose to protect their most valuable assets.

Less well known is that a Japanese midget submarine was spotted at 3:42am, four hours before the attack began. A destroyer, the USS *Ward*, was called in which failed to find that sub, but did find and sink a second sub at 6:37am, still more than an hour before the air strike. The *Ward* radioed in "We have attacked, fired upon, and dropped depth charges on a submarine operating in defensive sea areas." Would not this action have put the Fleet on high alert, unless someone overruled it?

At 7:02am, a full 53 minutes before the first bomb fell, radar operators at Opana Point detected the incoming Japanese aircraft. They alerted their superior, Lt. Kermit Tyler, who failed to make any report, but did however take his men away from their posts and to breakfast. Tyler's lack of action has long been considered suspicious by the conspiracy theorists.

Indeed, nearly a full year before the attack, the Commander-in-chief of the Pacific Fleet, Admiral Husband Kimmel, wrote to Washington:

> *I feel that a surprise attack (submarine, air, or combined) on Pearl Harbor is a possibility, and we are taking immediate practical steps to minimize the damage inflicted and to ensure that the attacking force will pay.*

Then, ten days before the attack, Kimmel was ordered to make just such a defensive deployment of the Fleet. And yet, on that morning, the ships were sitting ducks at their berths, the men asleep in their bunks, and most of the American aircraft were parked on the fields in plain view, packed into tight bunches, as if to deliberately make easy targets. It's also been pointed out that since the ships were sunk in the harbor, most were raised and repaired. Had they been sunk at sea they would have been lost. If you wanted to be attacked, but also wanted to be able to bounce back, this was the way to do it.

Combined with the fact that the Americans had broken the primary Japanese diplomatic code called Purple and made some progress breaking the military code JN-25, and had access to some Japanese intelligence, it seems hard to reach any conclusion other than the United States knew the attack was coming and deliberately allowed it to happen.

Or, at least, so we might conclude if we considered only the above points. But it turns out that if we examine each of these points not just with a narrow focus to see only the suspicious side, and look at the complete event in context, no good arguments for a conspiracy remain. Most of the points made by conspiracy theorists were raised by the 2000 book *Day of Deceit* by Robert Stinnett, who really boiled down the innuendo from the preceding 59 years and condensed it into a cohesive conspiracy. However, it should be noted that many more authors (almost all others) find him to be wrong. Chief among these is probably *Pearl Harbor: Final Judgment* from 1992, written by Henry Clausen. In 1944, the Secretary of War ordered Clausen, then a lawyer in the U.S. Judge Advocate's office, to conduct an independent investigation into what really happened in the days and months leading up to Pearl Harbor, and to find out who screwed up. His report remained top secret until its substance was finally published in this book.

Clausen found plenty of sloppiness, but nothing that could be characterized as a cool, smoothly-running conspiracy. Agencies operated independently, decoding Japanese transmissions and then filing them away rather than sharing them. There was plenty of knowledge that hostility was building, but no experience in how to deal with it and no specific knowledge that it was so imminent. Roosevelt knew as much as anyone, and issued warnings and ordered preparations that were poorly handled all the way down the line.

One thing that conspiracy theorists and historians agree upon is that Admiral Kimmel was unjustly made the scapegoat for Pearl Harbor. Ten days after the attack, he was reduced in rank and replaced by Admiral Nimitz. It's also agreed that he did the best he could given the limited amount of intelligence Washington shared with him, and this is one point where the conspiracy theory simultaneously kicks in and breaks down. Historians say he was held accountable for bad decisions; conspiracy theorists say he was made the scapegoat for the secret orders from Washington. But, nearly everything that happened at Pearl Harbor was on Kimmel's *own* orders. Let's look at some.

When Kimmel received the order to assume defensive positions ten days before the attack, viable threats at the time were from espionage and sabotage, not actual attacks. Thus the aircraft were moved out into the open and tightly packed, where they could be best guarded against saboteurs. The ships were similarly grouped in the harbor. It was the wrong interpretation of the order, but it was a reasonable one in the context of what Kimmel knew was happening.

Admiral Husband Kimmel

How true is it that the three carriers were safely hidden out at sea? Not very. The carriers were not clustered safely together; they were widely scattered throughout the Pacific on separate duties. Being alone out at sea even with their carrier groups, each isolated far away and unable to support one another, was not at all considered safe. The USS *Saratoga* was just coming out of a lengthy overhaul in Seattle and was underway to Pearl Harbor via San Diego at the time of the attack, but the USS *Enterprise* and the USS *Lexington* had in fact been at Pearl and recently sent away. Why?

Kimmel had sent them, separately and on staggered schedules, to deliver Army aircraft to reinforce Midway and Wake islands. Because of the Japanese spy network on Hawaii, great caution was taken to disguise this movement of forces. The *Enterprise* was scheduled to return by December 5th, at which time the *Lexington* would leave; Kimmel wanted to make sure that Pearl had coverage from at least one carrier at all times. The *Lexington* left on schedule, but unfortunately, bad weather struck the *Enterprise* and kept its group at sea for two extra days, resulting in an unforeseeable 2-day span of no carriers in Pearl Harbor. There was never any mysterious directive from Washington to hide the carriers. Had the weather not intervened, there would have been at least one carrier in Pearl at all times, which was the maximum force available.

Even so, there's a powerful reason why the absence of carriers would not support a conspiracy theory. World War II was the first time that aircraft carriers proved themselves to be the most important assets in naval warfare. At the time of Pearl Harbor, we'd not yet learned that, and the battleship was considered the most crucial weapon. That's why the Pacific Fleet had nine battleships and only three carriers. Conspiracy theorist descriptions of the battleships as old, useless, and expendable are a misstatement of history. They were the best we had, and their perceived value was such that at the time of the Pearl Harbor attack, six new battleships of the Iowa class were under construction, and a further six of the Montana class were planned. It wasn't until the Battle of Midway in 1942 that we learned the value of carriers, and construction shifted to those.

Was the *Ward's* sinking of the submarine covered up to prevent an alarm? The *Ward's* report made it to the desk of the watch officer at 7:15. At 7:30, Kimmel and Rear Admiral Claude Bloch both received it separately by telephone. By the time the Japanese attacked, 25 minutes later, Kimmel and Bloch were still conversing to determine the significance of the sub incident. Kimmel's opinion was that this was probably one more in a long line of false reports of submarines they'd been accustomed to receiving. Five minutes before the air strike, Kimmel ordered the destroyer USS *Monaghan* to go and verify the *Ward's* story. The *Monaghan* never made it. Kimmel's hunch was only conclusively proven wrong in 2002, when the midget submarine's wreck was discovered.

When Opana Point picked up the Japanese attack force on radar, their station was still under construction and was not yet fully operational. It had been staffed but nobody had yet received any training. The serviceman at the scope had, in fact, never used the equipment before at all. Lt. Tyler was a fighter pilot, and this was only his second day at Opana Point, and he had not been trained yet either. When Tyler was informed of the inbound target, he assumed it to be a flight of B-17's known to be inbound on that same course, which was a pretty common event. Since nobody perceived that anything unusual was happening, Tyler famously said "Don't worry about it," and they did in fact all go to breakfast. But once the attack began, they ran on foot back to the radar station and helped as best they could. A 1942 court of inquiry cleared Tyler of any blame, and he went on to have an exceptional career in the Air Force.

Now of course, all this only pertains to what happened at Pearl Harbor on the day of the attack. It doesn't address the much larger question of what President Roosevelt might have known or wanted to happen, or other people in Washington. The reason I don't go into that is that it doesn't matter. Even if this presumed conspiracy to allow the attack did exist, it failed to have any effect where the rubber meets the road. No orders from Washington altered the state of readiness at Pearl Harbor. Obviously the attack ultimately did play into the hands of anyone who wanted war with Japan; every tragedy somehow benefits somebody. That doesn't make every tragedy a conspiracy.

References & Further Reading

Borch, F., Martinez, D. *Kimmel, Short, and Pearl Harbor: The Final Report Revealed.* Annapolis: Naval Institute Press, 2005. 66.

Brewer, B. "The Missing Carriers of Pearl Harbor." *Times-Herald.* 28 Sep. 1944, Newspaper.

CIA. "Intelligence at Pearl Harbor." *FOIA Electronic Reading Room.* Central Intelligence Agency, 22 Aug. 1946. Web. 6 Jun. 2010. <http://www.foia.cia.gov/docs/DOC_0000188601/0000188601_0009.gif>

Clausen, H., Lee, B. *Pearl Harbor: Final Judgement.* New York: Crown, 1992.

Outerbridge, W. "USS Ward, Report of Pearl Harbor Attack." *Naval Historical Center.* United States Navy, 13 Dec. 1941. Web. 11 Jun. 2010. <http://www.history.navy.mil/docs/wwii/pearl/ph97.htm>

Stinnett, R. *Day of Deceit: The Truth About FDR and Pearl Harbor.* New York: Free Press, 2000.

Various. "USS Ward Crew: In Their Own Words." *USS Ward DD-139.* SpecWarNet, 16 Nov. 2002. Web. 10 Jun. 2010. <http://www.specwarnet.net/USSWard/crew.htm>

29. Mozart and Salieri

Was Wolfgang Amadeus Mozart actually murdered by his rival composer Antonio Salieri?

The legend first entered the public consciousness, in a significant way, with the 1984 movie *Amadeus*. In it, Wolfgang Amadeus Mozart was killed by his jealous rival, the court composer Antonio Salieri. Salieri cleverly took advantage of Mozart's fondness for drink, his financial crisis, and his obsession with pleasing his deceased father, and tricked Mozart into working himself to death. He did this by anonymously commissioning a requiem mass against an impossible deadline, presumably for his own father, until the hapless Mozart simply collapsed under all the various pressures. Salieri then took the manuscript and published it as his own work, while Mozart's body was thrown into a pauper's grave. The movie won eight Oscars including Best Picture and numerous other awards worldwide; and if you ask the average person on the street how Mozart died, the story from the movie is the one they'll probably tell.

But how true is it? Interestingly, although it is entirely fictionalized, there may actually be a grain of fact that's larger than you might think.

The movie was based on Peter Shaffer's 1979 stage play *Amadeus*. Like the movie, the story is told in flashback, with Salieri giving his confession to a priest after a failed, guilt-driven suicide attempt. As his own music fades into obscurity and Mozart's grows more and more popular even after his death, Salieri makes a final attempt to be relevant in the public's eye. He leaves a false confession that he poisoned Mozart with arsenic. But ironically, nobody believes his confession, leaving Salieri more marginalized than ever.

Shaffer's play was in turn inspired by the great Russian poet and author Alexander Pushkin, who wrote a short drama called *Mozart and Salieri* in 1830, only five years after Salieri's death. In 1898, Pushkin's piece was also used almost word-for-word as the libretto for a one-act opera of the same name by Nikolai Rimsky-Korsakov. In this darker and much simpler tale, Salieri, jealous of Mozart's skill as a composer and resentful of his low character, invites Mozart to dinner. They play almost a cat-and-mouse game until Salieri finally gets a chance to pour poison into Mozart's drink. Salieri's celebratory song hints at a descent into insanity.

Unfortunately, there will be no unmasking of the true perpetrator today. The cause of Mozart's death in 1791 is as much debated today as it was then; and if poison was indeed the instrument, there is anything but a consensus on who might have been the assassin.

History makes no secret of the fact that Mozart and Salieri were professional rivals. During their years together in Vienna, Salieri was greatly respected professionally. Emperor Joseph II liked him a lot, and Salieri held successive roles as court composer, director of Italian opera, and court conductor. Mozart and their mutual friends often spoke openly of Salieri's efforts to influence the availability of theaters and performers to favor his own shows at the expense of Mozart's. Salieri owned Italian opera in Vienna, and Mozart's forays into the genre were a clear step on Salieri's toes. Other composers in other genres had no such problems with Salieri, and neither did Mozart when he avoided Italian opera. There was no need for Salieri to kill Mozart to get him out of the way; Salieri's position in the industry gave him all the power he needed.

At the core of the question of Salieri's guilt is an enduring legend that he gave a deathbed confession. In 1823, some 32 years after Mozart's death, Salieri did indeed make a failed suicide attempt by cutting his own throat. Rumors quickly spread that he had confessed to killing Mozart, and these rumors became so widespread and persistent that a leaflet was distributed at a Vienna performance of Beethoven's Ninth Symphony depicting Salieri standing over Mozart with a cup of poison. Mozart's music was widely loved by this time, and with his alleged killer still alive, public fever over the murder ran high. But the hatred of Salieri was not unanimous. Two camps formed: Those who knew Salieri and defended him against the rumors, and the much larger Court of Public Opinion composed largely of people who had not known either man and had little firsthand know- ledge.

Antonio Salieri

No reliable evidence exists that Salieri ever made such a confession. The best anyone's come up with is described in a pamphlet published in Moscow in 1953, wherein author Igor Boelza claims he was told by another man (who had since died) that he'd once seen a report of the confession written by Salieri's priest. He also alleged that the deceased man showed the report to a number of academics, whom he fails to name. No such report is known to exist — which would be a huge discovery to any academic who had actually seen it — so judge the reliability of Boelza's pamphlet for yourself. What does exist is a written statement from two men who were Salieri's 24-hour caregivers during the last two years of his life, stating that they never heard him make any such confession.

There is also an anecdote that Salieri once took the very young composer Rossini to meet Beethoven at his home in Vienna. Beethoven allegedly turned Rossini away and shouted "How dare you come to my house with Mozart's poisoner?" Although this story is often repeated, it's inconsistent with what was known of Beethoven and Salieri. Salieri had tutored Beethoven, and the two had always been friends. Beethoven held his tutor in such high esteem that, even after Mozart's death, he dedicated his violin sonatas Opus 12 to Salieri, and wrote a series of variations on a theme from Salieri's opera *Falstaff*. So even this anecdote seems unreliable.

Defending Salieri against the rumors of murder were all of Mozart's principal biographers, and Salieri's other pupils, which included Mozart's closest confidant, his own student Süssmayr, who completed Mozart's Requiem after his death. Süssmayr was with Mozart nearly daily throughout his final months, and knowing the causes of Mozart's illness as well as anyone, he did not hesitate to continue his studies under Salieri. It's also noteworthy that Salieri was never under any kind of official suspicion of criminal activity. Indeed, his professional career continued to flourish despite the rumors. Many great composers continued studying under him, including the young Franz Liszt and Franz Schubert.

Although it was Salieri who took the heat for Mozart's alleged murder, he was not the only suspect. In contrast to the popular legend, Salieri was not even the one who commissioned the Requiem upon which Mozart forced himself to work so hard even until the day of his death; that patron was Count Franz von Walsegg, who wanted the Requiem to honor his late wife.

A number of authors have put forward the hypothesis that Mozart, who was a Freemason, was killed by a Masonic conspiracy. Why would the Freemasons murder one of their own? The most often cited reason is that Mozart's opera *The Magic Flute* somehow violated Masonic rules. One claim is that the story conceals an allegory for an alleged plot to overthrow

Freemasonry; another is that it contained misuses of Masonic symbols. Author Georg Friedrich Daumer was the most vocal proponent of these theories, which he first published in 1861. However, his belief that Freemasons poisoned Mozart should be viewed in the context of his other claims: He also believed that Freemason conspiracies murdered many heads of state and leaders in religion and philosophy.

But even the very idea that *anyone* was responsible for Mozart's death is not generally accepted among modern historians. The most thorough accounts of Mozart's four months of illness all come from his wife, Constanze, and from shorter reports from the friends and associates who frequently visited him, including her sister Sophie. None thought he had been poisoned. Several times, Mozart told Constanze that he believed he had been poisoned with a popularly known arsenic-based potion called *aqua tofana*, however he dismissed the notion himself during a spell in which his health seemed to return for a time. His principal symptom was swelling, particularly of the extremities, which caused him great pain when it was at its worst. At the application of a cold compress to his forehead on December 5, 1791, the shock caused him to lose consciousness, from which he never awoke, and died two hours later. Mozart's own doctors blamed his death on "high miliary fever", but this was a prescientific diagnosis and does not correspond to any specific diseases now known.

After his death, Mozart's first biography was written by Franz Niemetschek and was based on interviews with Constanze and Sophie and numerous documents provided by them. His second biography was written by Constanze's second husband, Georg Nikolaus von Nissen. Neither book suggests that Mozart died from any cause other than illness. Together, these two works comprise the most detailed reports available of Mozart's physical condition, and it's from these books that modern doctors have tried to theorize what disease he might have had.

The prevailing theory is chronic kidney disease. Mozart was probably at high risk of this anyway; as a child he'd been ill with what's now believed to be scarlet fever and rheumatic fever, both of which can cause kidney damage. This diagnosis is consistent with the reports of Mozart's condition. Kidney failure would have caused the type of edema, or swelling, that was reported, and the uremic poisoning would have ultimately killed him. But some have noted that people so afflicted are often unable to work at all or even comatose, while Mozart was doing some of his best work during his final months.

Mozart's self-diagnosis of arsenic poisoning from *aqua tofana* seems unlikely. Although that poison was easily obtainable (it was actually in commercial production and sold as a murder weapon), arsenic poisoning does not produce any of the symptoms which afflicted Mozart.

Some doctors have also speculated that he could have died of mercury poisoning, which may or may not have been given to Mozart by someone else. Mozart may have even been administering mercury to himself as a cure for syphilis — a treatment that some physicians of the day were promoting. It's unlikely that the true cause of Mozart's death will ever be known. It has never been possible to exhume Mozart's body for testing because he was, as depicted in the movie, buried in a mass, unmarked grave.

It was a small service on a cold morning. Mozart and Constanze had both expressed their dislike of the pomp and ceremony of funerals, so they chose a burial in a style that had actually been a law under Joseph II until it was repealed just a few years before Mozart's death, and that was to dispense with caskets and lavish services in favor of simple burials with the body sewn into a cloth sack. Constanze herself was too bereaved and chose not to attend, and so only a very few close friends and family walked with the body. They stopped at the cemetery gates and bid their last farewells. Mozart was taken alone to his final resting spot, and the mourners turned away. Among this small group dressed in black, this tightest circle of those who were the last to be with him, was Antonio Salieri.

References & Further Reading

Boerner, S. "Biography." *The Mozart Project.* MozartProject.org, 25 Apr. 1998. Web. 14 Jun. 2010.
<http://www.mozartproject.org/biography/>

Borowitz, A,. "Salieri and the 'Murder' of Mozart." *Musical Quarterly.* 1 Apr. 1973, Volume 59, Number 2: 263-284.

Deutsch, O. *Mozart: A Documentary Biography.* Stanford: Stanford University Press, 1965.

Niemetschek, F. *Life of Mozart.* Westport: Hyperion Press, 1979.

Nissen, G. *Biographie W. A. Mozart's.* Leipzig: Gedruckt, 1828.

Stafford, W. *The Mozart Myths: A Critical Reassessment.* Stanford: Stanford University Press, 1991.

30. The Things We Eat...

Do you really know which kinds of food are good for you, and which are bad?

In this chapter, we're going to take a collective look at all the conflicting warnings and exhortations we hear about what we should and shouldn't eat. It seems everyone has some pet theory that you shouldn't drink milk, or you have to eat organic, or you shouldn't eat "processed" foods, or you must only eat raw. There are always explanations for why this is: We didn't "evolve" to eat this or that; it isn't "natural" to eat something; our digestive systems weren't meant to handle a certain thing. I know what you're thinking: How is it possible to cover all those possible claims in a single short book chapter? We're going to do it by stepping back from all of the specific claims and specific foods, way back. We're going to look at food as a whole, and study what it's made of, what those bits are, see what we need and what we don't. And then, with this as a foundation, we'll have the tools to effectively examine any given eating philosophy.

Originally, this chapter was going to be about the specific claim that we shouldn't drink milk, based on the idea that humans are the only species that drinks another species' milk, and it's therefore unnatural. I've also been given the suggestion — several times — that we should never give pet food to pets, because its ingredients are not the ones they evolved to eat. I quickly realized that all of these notions are basically the same, and all depend on a fundamental misunderstanding of the nature of food. Dog food, beer, cheese, and cake frosting are all compounds that no species evolved to eat. Then how is it that we're able to eat them? In essence, it's because all food — in whatever strange form we want to present it — consists of the same basic building blocks, all of which we *did* evolve to eat, and all of which are found in nature.

Before we look at these building blocks, I need to state that it's impossible to be 100% comprehensive within the limitations of these few pages. There are innumerable subtleties and exceptions and footnotes that I'm not going to go into. Most of these exceptions come from the fact that humans developed in a broad range of environments, and as a result, some groups are more or less adapted to certain compounds, lactose tolerance being an obvious example. People with phenylketonuria can't metabolize the amino acid phenylalanine. Some populations have difficulty synthesizing enough Vitamin D in their skin. These are just examples; there are plenty of others, and I'm not pretending to cover every nuance here. If you want to delve further, see the Further Reading suggestions at the end of the chapter. Today's discussion is at a level that applies generally to all humans, and to some degree to most other vertebrates as well.

Food breaks down into six basic compounds. All food consists of combinations of these six, and every one of them is found in nature:

1. Amino Acids

These are the building blocks of proteins. Proteins are essential for our bodies. We need to eat protein, which is then broken down by our digestive system into its constituent amino acids, and then our body reassembles them into whatever proteins it needs. Some amino acids are called essential, and this refers to those that our body cannot synthesize and that we must eat. There are eight essential amino acids, plus about fourteen others that are conditionally essential: needed by infants, growing children, and other certain populations. With few exceptions, the body makes use of all amino acids; there's no such thing as an amino acid that we can't or shouldn't consume. Proteins in food like enzymes and hormones are usually not used by the body as enzymes and hormones; they too are broken down into amino acids which are then gainfully employed as building blocks.

2. Fatty Acids

Like amino acids, fatty acids come in essential and conditionally essential varieties. Omega-3 and omega-6 are the two essential fatty acids that we must get from food because we can't synthesize them, and that have a wide range of important functions throughout our bodies; three others are usually considered conditionally essential for some populations.

All the rest of the fatty acids are ones that we don't need to eat. Our body does usefully employ most of them, but it can synthesize what it needs, so you generally want to minimize your food intake of them. These include saturated fats (where all available chemical bonds are "saturated" with a hydrogen atom) and the non-essential unsaturated fats, which include monounsaturated, polyunsaturated, and trans fats.

3. Carbohydrates

These are your sugars and starches, which all break down into monosaccharides: the single sugars glucose, fructose, galactose, xylose, and ribose. Two of those together may come from a disaccharide like table sugar; a longer polysaccharide chain may come from the carbs in a granola bar. Whatever we eat gets broken down into those monosaccharides (though some populations may have enzymatic deficiencies that hamper the digestion of some combinations, like lactose). Those monosaccharides fuel our metabolism, and are the principal building blocks of the synthesis of other needed compounds. Any extra monosaccharides are put together into space-saving polysaccharides for storage.

4. Vitamins

Exactly what is a vitamin? There's a simple and clear definition. We've just discussed the three basic types of nutrients; a vitamin is any other organic compound that our body needs, that we are unable to synthesize enough of, and that we must get from food. Vitamins were discovered throughout the first half of the 1900's, and each time we learned about a new one, it was given a successive identifying letter: Vitamin A, B, C, and so on. After we learned about Vitamin B we found it was actually eight different vitamins, and so we have Vitamin B_1, B_2, B_3, and the rest. Many animals synthesize these vitamins from proteins and fats, so they don't need to eat such a diversity of different foods to get them, the way we do.

There are two basic kinds of vitamins: water soluble (vitamins B and C) and fat soluble (all the others). If you consume more water soluble vitamins than you need, the excess will be quickly and harmlessly discharged in your urine. Overdosing on fat soluble vitamins provides a bit more of a challenge to your body though, and can lead to hypervitaminosis, which can be dangerous in extreme cases.

With a few notable exceptions, anybody who lives and eats in a modern industrialized country gets more than enough of all the vitamins their body needs, and there's no need to spend money on vitamin supplements. If you eat three meals a day, the buckets in which your body has room to store vitamins are brim full, and vitamin supplementation would be like pouring more onto an already overflowing bucket. Save your money.

5. Minerals

These are defined as the inorganic chemical elements that our body needs. There are sixteen essential elements (chemically, they're not really all minerals) including iron, calcium, zinc, sodium, and potassium. There are some half-dozen others considered conditionally essential, but if you stick with the sixteen you're probably all right. Minerals obviously have to be consumed; our bodies are not atomic reactors and so we can't synthesize chemical elements.

With a very few exceptions, anyone who eats regular meals in an industrialized country gets more than enough of all the minerals they need. Perhaps the two most common exceptions are pregnant women who can benefit from iron supplementation, and people who avoid dairy products and could often benefit from calcium supplementation.

6. Water

Kind of an obvious one. It's the only thing anyone needs to drink — there's no substitute — and most of us get all we need from what's contained in our food and other drinks.

And so, there we have the six fundamental compounds that make up all food. The basic argument against all of the various "You shouldn't eat this or that" claims is that those foods all break down into the same building blocks, building blocks which you would also get from other food. The opposing argument in favor of those claims is that some of these building blocks are good (like essential amino acids) and some are bad (like trans-fat), and we should strive to eat foods that deliver the most good nutrients with the least amount of harmful contents. Kind of a no-brainer, obviously, but it's rarely the argument that's actually made. Instead, the arguments I usually hear call out a particular food based on some ideology rather than its actual contents. Not that there's anything wrong with ideologies, but they should not be misrepresented as food science.

Other than a glass of pure water, there is hardly a food source on the planet that delivers anything less than a radically complex assortment of proteins, lipids, and starches, laced with vitamins and minerals. It's the proportions that differ. Looking at it from this perspective, there's little fundamental difference between milk and orange juice. The orange juice contains more sugar and vitamins but less fat and protein, while the milk contains a more even spectrum of nutrients. An argument like "Cow's milk is bad because early humans didn't evolve to drink it" becomes completely goofy when you consider only this one irrelevant characteristic. The same goes for arguments against manufactured pet food. There is no reason at all why pet food should look like, or come from the same source as, the animal's natural food; so long as it delivers the nutrients the animal needs.

Cooking introduces chemical changes that are, for the most part, the same as the first step in digestion. Some compounds cannot be digested unless they're cooked first to break certain chemical bonds. Most claims that cooking destroys nutrients are wrong; cooking merely starts the ball rolling on what your digestive system was going to do to the food anyway.

One nice thing about being a technological society is that we have the capability to understand food science, and to design nutritious foods that are more attractive and tasty than our ancestors were able to find on the savannah. The bottom line is that if you wish to evaluate any given food's nutritional value, you must look at what it actually delivers. Simply considering where it came from, or who designed it, is not a useful assessment of its actual substance.

References & Further Reading

ADA. "Position of the American Dietetic Association: Nutrient Supplementation." *Journal of the American Dietetic Association.* 1 Dec. 2009, Volume 109, Issue 12: 2073-2085.

Chiras, D. *Human Biology.* Sudbury: Jones & Bartlett Publishers, 2005. 81-92.

Holick, M. "Vitamin D Deficiency." *New England Journal of Medicine.* 19 Jul. 2007, Volume 357, Number 3: 266-281.

Kennedo, G. "Dietary Reference Intakes Tables and Application." *Institute of Medicine.* National Academies of Sciences, 14 Jan. 2010. Web. 25 Jul. 2010.
<http://www.iom.edu/Activities/Nutrition/SummaryDRIs/DRI-Tables.aspx>

Simopoulos, A., Cleland, L. *Omega-6/omega-3 essential fatty acid ratio: the scientific evidence.* Basel: S. Karger AG, 2003.

USDA. "USDA Nutrition Evidence Library, 2010." *Nutrition Evidence Library.* USDA Center for Nutrition Policy and Promotion, 15 Jun. 2010. Web. 26 Jul. 2010.
<http://www.nutritionevidencelibrary.com/default.cfm?>

USDA. "Questions To Ask Before Taking Vitamin and Mineral Supplements." *Nutrition.gov.* USDA National Agricultural Library, 11 Jun. 2009. Web. 7 Jul. 2010.
<http://www.nutrition.gov/nal_display/index.php?info_center=11&tax_level=2&topic_id=1939>

31. Some New Logical Fallacies

A look at some newer logical fallacies, often used in place of sound arguments.

One of the most popular Skeptoid podcast episodes ever was *A Magical Journey through the Land of Logical Fallacies* (see Volume 2 of this book series). In it, we looked at some of the most common fallacious ways to argue a point; in essence, the use of rhetoric as a substitute for good evidence. Logical fallacies can be deliberately employed when you don't have anything real to support the point you want to make, and they can also be accidentally employed when you mistake compelling rhetoric for a sound argument. Good attorneys and debaters are experts with wielding fallacious logic, as are the most successful salespeople of quack products.

In the adventure of producing Skeptoid, I'm frequently deluged by logical fallacies in emails from those who disagree with me. On the Skeptalk email discussion list, we often have fun identifying such fallacies in news articles or promotions by charlatans. As a result of all this experience, I've compiled a list of some newer logical fallacies we've found most entertaining. Now, admittedly, some of these are pretty similar to the traditional fallacies, but you may be more likely to recognize them in their contemporary guise. Let's begin with:

Appeal to Lack of Authority

Authority has a reputation for being corrupt and inflexible, and this stereotype has been leveraged by some who assert that their own lack of authority somehow makes them a better authority.

Starling might say of the 9/11 attacks: "Every reputable structural engineer understands how fire caused the Twin Towers to collapse."

Bombo can reply: "I'm not an expert in engineering or anything, I'm just a regular guy asking questions."

Starling: "We should listen to what the people who know what they're talking about have to say."

Bombo: "Someone needs to stand up to these experts."

The idea that not knowing what you're talking about somehow makes you heroic or more reliable is incorrect. More likely, your lack of expertise simply makes you wrong.

Proof by Anecdote

Many people believe that their own experience trumps scientific evidence, and that merely relating that experience is sufficient to prove a given claim.

Starling: "Every scientific test of magical energy bracelets shows that they have no effect whatsoever."

Bombo: "But they work for me, therefore I know for a fact they're valid and that science is wrong."

Is Bombo's analysis of his own experience wrong? If it disagrees with well-performed controlled testing, then yes, he probably is wrong. Personal experiences are subject to influences, biases, preconceived notions, random variances, and are uncontrolled. Relating an anecdotal experience proves nothing.

Michael Jordan Fallacy

This one can be used to impugn the motives of anyone in the world, in an effort to prove they are driven by greed and don't care about anyone else's problems:

> *Bombo:* "*Just think if Michael Jordan had used all his talents and wealth to feed third world children, rather than to play a sport.*"

Of course, you can say this about anyone, famous or not:

> *Bombo:* "*If your doctor really cared about people's health, he'd sell everything he owned and become a charitable frontier doctor in Africa.*"

In fact, for charitable efforts to exist, we need the Michael Jordans of the world playing basketball. Regular non-charitable activities, like your doctor's business office, are what drives the economic machine that funds charity work. The world's largest giver, the Bill & Melinda Gates Foundation, would not exist had a certain young man put his talents toward the Peace Corps instead of founding a profitable software giant.

Proof by Lack of Evidence

This one is big in the conspiracy theory world: The lack of evidence that would support their conspiracy theory is due to the evil coverup. Thus, the lack of evidence for the conspiracy is, in and of itself, evidence of the conspiracy.

> *Bombo:* "*The passengers on Flight 93 were taken off the plane and executed by the government.*"
>
> *Starling:* "*But there's no evidence of that.*"
>
> *Bombo:* "*Exactly. That's how we know it for a fact.*"

There are certainly things in the world that are true but for which no evidence exists, but these are in the minority. If you want to be right more often than not, stick with what we can actually learn. If instead your standard is that anything that can't be disproven must therefore be true, like Russell's teapot, you're one step away from delusional paranoia.

Appeal to Quantum Physics

This is a form of special pleading, a scientific-sounding way of claiming that the way your magical product or service works is beyond the customer's understanding; in this case, based on quantum physics. That sounds impressive, and who's qualified to argue? Certainly not the average layperson.

> *Bombo: "Quantum physics explains why pressure points on the sole of your foot correspond with other parts of your anatomy."*

Here's a tip. If you see or hear the phrase "quantum physics" mentioned in a context that is anything other than a scientific discussion of subatomic theory, raise your red flag. Someone is probably trying to hoodwink you by namedropping a science that they probably understand no better than your cat does.

Proof by Mommy Instinct

Made famous by antivaccine activist Jenny McCarthy, this one asserts that nobody understands health issues better than a mom. Mothers obviously have experience with childbirth and with raising children, but is there any reason to suspect they understand internal medicine (for example) better than educated doctors, many of whom are also mothers? Not so far as I am able to divine.

Remember that Mommy Instincts are no different than anecdotal experiences. They are driven by perception and presumption, not by science.

Argument from Anomaly

This one is big with ghost hunters and UFO enthusiasts. Anything that's anomalous, or otherwise not immediately, absolutely, positively, specifically identifiable, automatically becomes evidence of the paranormal claim.

> *Starling: "We found a cold spot in the room with no apparent source."*
>
> *Bombo: "That must be a ghost."*

Since the anomaly is, well, an anomaly, that means (by definition) that you can't prove it was anything other than a ghost or a UFO or a leprechaun or whatever they want to say. Since the skeptic can't prove otherwise, the Argument from Anomaly is a perfect way to prove the existence of ghosts. Or, nearly perfect, I should say, because it's not.

Chemical Fallacy

Want to terrify people and frighten them away from some product or technology that you don't like? Mention chemicals. Chemical farming, chemical medicines, chemical toxins. As scary as the word is, it's almost meaningless, because everything is a chemical. Even happy flowers and kittens consist entirely of chemicals. It's a weasel word, nothing more, and its use often indicates that its user was unable to find a cogent argument.

Appeal to Hitler

This one is inspired by Godwin's Law, in which Mike Godwin stated "As an online discussion grows longer, the probability of a comparison involving Nazis or Hitler approaches 1." Ever since, such arguments have become known as the *reductio ad Hitlerum*, or the Appeal to Hitler. It's a garden variety "guilt by association" charge, saying you're wrong because Hitler may have thought or done something similar.

> *Bombo: "You think illegal aliens should be deported? Sounds exactly like how the Nazis got started."*

Starling gives the common reply:

> *Starling: "The Nazis also owned dogs and played with their children."*

For good measure, Bombo comes back with a "straw man on a slippery slope" argument:

> Bombo: *"Are you saying everything about the Nazis was perfect?"*

Proof by Victimization

Beware of claims from those lording their victimization over you. They may well have been victimized by something, be it an illness, a scam, even their own flawed interpretation of an experience. And in many cases, such a tragedy does give the victim insight that others wouldn't have. But it doesn't mean that person necessarily understands what happened or why it happened, and should not be taken as proof that they do.

> Bombo: *"My neighbor's Wi-Fi network gave me chronic fatigue."*
>
> Starling: *"But that's been disproven every time it's been tested."*
>
> Bombo: *"You don't know what you're talking about; it didn't happen to you."*

Victimization does not anoint anyone with unassailable authority on their particular subject.

Better Journal Fallacy

It's common for purveyors of woo to trot out some worthless, credulous magazine that promotes their belief, and refer to it as a peer-reviewed scientific journal:

> Starling: *"If telekinesis was real, you'd think there would be an article about it in the American Journal of Psychiatry."*
>
> Bombo: *"That rag is part of the establishment conspiracy to suppress psi research. You need to turn to a reputable source like the Journal of the American Society for Psychical Research. It's peer-reviewed."*

And so it is, but its reviewers are people who have failed to establish credibility for themselves, as have such journals themselves. There are actually metrics for these things. The productivity and impact of individual researchers can be described by their Hirsch index (or *h*-index), which attempts to measure the number and quality of citations of their publications and research. A journal's reputation can be shown by its impact factor, which measures approximately the same thing. Although these indexes are not perfect, you need not ever lose a "my peer-reviewed scientific journal is better than yours" debate. Look up impact factors in the Thomson Reuters *Journal Citation Reports* through sciencewatch.com.

Appeal to Dead Puppies

Sometimes tugging at the heartstrings with a tragic tale is enough to quash dissent. Who wants to take the side of whatever malevolent force might be associated with death and suffering?

Starling: "Thank you, door-to-door solicitor, but I choose not to purchase your magazine subscription."

Bombo: "But then I'll be forced to turn to drugs and gangs."

Oh no! What a horrible image. The Appeal to Dead Puppies draws a pathetic, poignant picture in order to play on your emotions. Recognize it when you hear it, and keep your emotions separate from the facts.

Add these new fallacies to your arsenal. And remember to keep an eye out for them: The spotting of logical fallacies in pop culture can be a fun game, like looking for state license plates on the freeway. Learning to spot them also sharpens your critical thinking skills, so be on the lookout.

References & Further Reading

Albrecht, K. *Brain Power: Learn to Improve Your Thinking Skills.* Englewood Cliffs: Prentice-Hall, 1980. 167-183.

Curtis, G. "What is a logical fallacy?" *The Fallacy Files*. Gary N. Curtis, 21 Feb. 2004. Web. 5 Sep. 2011.
<http://www.fallacyfiles.org/>

Gula, Robert J. *Nonsense: Red Herrings, Straw Men and Sacred Cows: How We Abuse Logic in Our Everyday Language.* Mount Jackson, Va: Axios Press, 2002.

Novella, S. "Top 20 Logical Fallacies." *Skeptics Guide to the Universe.* SGU Productions LLC, 8 Feb. 2009. Web. 5 Sep. 2011.
<http://www.theskepticsguide.org/resources/logicalfallacies.aspx>

Shuster, K., Meany, J. *On That Point!: An Introduction To Parliamentary Debate.* New York: IDEA, 2003. 313-315.

Whitman, G. "Logical Fallacies and the Art of Debate." *Glen Whitman's Home Page.* California State University, Northridge, 29 Jan. 2001. Web. 5 Sep. 2011.
<http://www.csun.edu/~dgw61315/fallacies.html>

32. The Astronauts and the Aliens

A close look at some of the stories of UFOs said to have been reported by NASA astronauts.

It was 1962 and American John Glenn was orbiting the Earth in *Friendship 7*, his capsule on the Mercury-Atlas 6 flight. Ground controllers were mystified at Glenn's report of fireflies outside his window, strange bright specks that clustered about his ship. The first thought was that they must be ice crystals from

Friendship 7's hydrogen peroxide attitude control rockets, but Glenn was unable to correlate their appearance with the use of the rockets. Astronauts on later flights reported similar bright specks, and eventually we learned enough about the space environment to identify what they were. Spacecraft tend to accumulate clouds of debris and contamination around themselves, and even though Glenn's rockets sprayed jets of crystals away from the capsule, many of the crystals would gather in this contamination cloud, where they reflected sunlight and interacted with other gases in the cloud. Experiments on board *Skylab* in the 1970's using quartz-crystal microbalances confirmed and further characterized this phenomenon. The case of John Glenn's mysterious fireflies was solved.

The stories of our humble explorations of the space around our planet tell of courage, danger, and adventure. But do they conceal another element as well? For as long as humans have had space programs, there have been darker tales flying alongside: tales of mysterious UFOs, apparently alien spacecraft monitoring our progress. These stories come from the early days of the Soviet launches, from the Mercury program, the Gemini program, the space shuttle flights, and perhaps most infamously from the Apollo flights to the moon.

Like pilots, astronauts are often given something of a pass whenever they report a UFO, a pass that presumes it's impossible for someone with flight training to misidentify anything they see in the sky. Most famously, Apollo 14 astronaut Edgar Mitchell, the sixth man to walk on the moon, has long maintained that most UFOs are alien spacecraft and that the government is covering up its ongoing active relations with alien cultures. Coming from a real astronaut, Mitchell's views are often quite convincing to the public.

NASA's reaction to Mitchell was anticlimactic, but highlighted that their business is launching things into space, not studying UFO reports:

> NASA does not track UFOs. NASA is not involved in any sort of cover up about alien life on this planet or anywhere in the universe. Dr. Mitchell is a great American, but we do not share his opinions on this issue.

Edgar Mitchell is a longtime proponent of psychic powers and alternate models of reality. During his Apollo flight he even conducted private ESP tests with his friends back home, and later went on to found the Institute of Noetic Sciences that researches telepathy and other such things. Mitchell does not claim to have personally observed any of these alien craft; he says his views are based on things told to him by people who are in on the secrets.

But other Apollo astronauts did see strange things. Perhaps the best known comes from Apollo 11, when Neil Armstrong, Buzz Aldrin, and Michael Collins noted that a UFO paced along with them for most of their flight to the moon. Depending on how carefully edited are the parts of the story you're given, this can sound like a most compelling event:

"We did watch a slow blinking light some substantial distance away from us," said Armstrong.

"There was something out there, close enough to be observed," said Aldrin:

> *"Now, obviously the three of us weren't going to blurt out, 'Hey, Houston, we've got something moving alongside of us and we don't know what it is,' you know? We weren't about to do that, because we knew that that those transmissions would be heard by all sorts of people and somebody might have demanded we turn back because of aliens or whatever the reason is."*

Documentaries have been made and books have been written promoting the idea that the episode was covered up by NASA or that the astronauts did not know what the mysterious visitor was, but in fact the period of uncertainty was quite brief. The astronauts and ground control were soon able to identify the object: It was one of four adapter panels that fit between the S-IVB third stage and the lunar module. As they were ejected from the S-IVB they were imparted with angular momentum, so as at least one panel continued along the same trajectory as the lunar module containing the crew, it rotated and blinked like a light as it reflected the sun.

As far as the episode being "covered up" by NASA? More likely, little official recognition was made simply because it was not a relevant part of the mission.

One famous picture was taken by the crew of Apollo 16 looking back as it was leaving the moon. Four seconds of video showed a near-perfect Hollywood flying saucer leaving some kind of plasma trail behind it. For over thirty years, a frame from this video was touted by the UFO community as proof of alien visitation. But then a team from Johnson Space Center's Image Science and Analysis Group decided to take it seriously. Without too much trouble, they found that when viewed from the window through which the video was taken, the object in the film was exactly consistent with a small floodlight protruding from the side of the capsule on a boom. The floodlight looked precisely like a flying saucer from that angle, and the boom matched perfectly with the plasma trail. Even a couple of bolts on the boom are visible in the video. You can see images from this analysis on NASA's web site.

Another report that's much touted by UFO enthusiasts came from James McDivitt, command pilot of Gemini 4 in 1965:

> "I was flying with Ed White. He was sleeping at the time so I don't have anybody to verify my story. We were drifting in space with the control engines shut down and all the instrumentation off (when) suddenly (an object) appeared in the window. It had a very definite shape -- a cylindrical object -- it was white -- it had a long arm that stuck out on the side. I don't know whether it was a very small object up close or a very large object a long ways away."

McDivitt took pictures, but was hampered by sun reflections on the window. There's a famously reproduced photograph described as "the tadpole", but according to McDivitt, that picture (selected by a NASA technician as the one he thought McDivitt was talking about) shows only the sun reflected on the window and is not what he saw.

In 1968, a study called the Condon Report was published. The Condon Committee, organized under physicist Edward Condon at the request of the US Air Force, studied UFO reports for the purpose of determining whether anything scientifically useful could be learned from them. McDivitt's report is famous in part because the Condon Report endorsed it as unidentified.

Common wisdom has always held that McDivitt's object was orbital debris from a rocket launch, either Soviet or American, even his own Titan II booster. One of Gemini 4's goals was to practice orbital rendezvous with the spent Titan II second stage, and though it corresponds closely with McDivitt's description of what he saw, McDivitt himself explains that he had spent two hours watching the Titan II stage as part of the exercise and was very familiar with its appearance. The object he saw, he insists, was not the booster he had grown to know so well.

Could McDivitt have been mistaken? Later in the same flight, Ed White radioed:

> *"We've got an object out in front of us. It's not flashing like it's the booster. It appears that it's that type of an object unless it's picking up some glow from the sun. It appears a very bright, very bright object... It was the booster. I can see the lights flashing on it now ... Just as it goes into darkness, the reflection of the sun on the booster causes a very bright image. That's the object I had seen earlier."*

McDivitt has never doubted that what he saw was merely an unmanned satellite or some other orbital debris. Detailed studies by others have found that the Titan II booster was in the right place and was the right distance away and was almost certainly the object. A Congressman inquired with NASA and was told "We believe it to be a rocket tank or spent second stage of a rocket." Only UFO enthusiasts who weren't there claim that it must have been an alien spacecraft.

Gemini 4 was not the only such flight where something similar happened, and that was also endorsed as unidentified by the Condon Report. Six months later, Gemini 7 with Frank Borman and Jim Lovell made what's been called a "football" maneuver to get them into a position where they would make recurring close approaches with their Titan II booster stage every orbit. When they did, Borman reported the following:

> *"Bogey at 10 o'clock high... We have several, looks like debris up here. Actual sighting... We also have the booster in sight..."*

> *Yeah, have a very, very many — look like hundreds of little particles banked on the left out about 3 to 4 miles... It looks like a path to the vehicle at 90 degrees... They are passed now — they were in polar orbit."*

Lovell reported the booster also in sight, "slowly tumbling", along with its associated debris cloud. So Borman's "bogey" particles had to be something other than the booster or booster debris: They were in a different direction, traveling along a different trajectory. Right?

Not necessarily. According to later analysis, Borman's bogeys were ice flakes from leftover fuel spewed from the Titan II, traveling along a parallel path to it. Imagine riding a bicycle beside two widely-separated train tracks. A train passing along the far track is so distant that it seems to be hardly moving; while a train passing suddenly along the near track seems many times faster. If your bicycle path is at the right angle, the train's relative movement may appear to be 90 degrees from your own. The ice particles could not have been on a polar orbit, as Borman speculated, because they would have passed at far too many miles per second to have been perceptible. Instead, the orbit of the bogey flakes and of the Titan II was only slightly offset from that of Gemini 7.

As often happens with such tales, popular retellings greatly exaggerate the event. Articles written about the Gemini 7 bogey have described it as "a massive spherical object", or even that Borman and Lovell "photographed twin oval-shaped UFOs with glowing undersides." These are nothing more than untrue embellishments added by UFO writers. A hoax photograph was even made to fit this latter description, and has been widely distributed.

Always remember that "unidentified" does not mean "positively identified as an alien spacecraft", something that UFO proponents forget all too often. There's a lot of stuff in orbit, and a lot of stuff traveling alongside every manned and unmanned spacecraft; and we'll always have UFO reports so long as we have a space program. While it may be intriguing to wonder which planet the Little Green Men came from, my experience is that the more fascinating science is that of trying to better understand what's *actually* happening.

References & Further Reading

Condon, E., University of Colorado. *Scientific study of unidentified flying objects.* Fort Belvoir: Defense Technical Information Center, 1969.

Hansen, J. *First Man: The Life of Neil A. Armstrong.* New York: Simon & Schuster, 2005. 430-432.

Morrison, D. "UFOs and Aliens in Space." *Skeptical Inquirer.* 1 Jan. 2009, Volume 33, Number 1: 30-31.

NASA. "SP-404 Skylab's Astronomy and Space Sciences." *NASA.* NASA History Division, Office of Communications, NASA, 29 Jan. 2002. Web. 5 Aug. 2010. <http://history.nasa.gov/SP-404/ch7.htm>

Oberg, J. *UFOs & Outer Space Mysteries: A Sympathetic Skeptic's Report.* Norfolk: Donning, 1982.

Petty, J. "UFO No Longer Unidentified." *NASA.* National Aeronautics and Space Administration, 19 Apr. 2004. Web. 3 Aug. 2010.
<http://www.nasa.gov/vision/space/travelinginspace/no_ufo.html>

33. Stalin's Human-Ape Hybrids

Did Josef Stalin order the creation of an army of half-ape, half-human hybrids, and did these experiments take place?

It was the Soviet dictator's dream: Soldiers with no fear, with superhuman strength and endurance, who would follow any order, eat anything, and ignore pain or injury. Workers who could do the labor of ten men without complaint, with no thought of personal time off, and no desire for pay. A force to carry the Soviet Union through its Five-Year Plan for economic development, and to make the nation invincible in war. Stalin's goal, according to modern mythology, was no less than a slave race of scientifically bred beings that were half human and half ape; a race he hoped would combine tremendous physical strength, dumb loyalty, and a human's ability to follow direction and perform complex tasks. But how much of this is true, and how much of it is the invention of modern writers and filmmakers looking for the sensational story?

It's no secret that a renowned Russian biologist, Il'ya Ivanovich Ivanov, spent much of his career working on just this. Around 1900 he gained great fame and national acclaim with his work on artificially inseminating horses, increasing the number of horses that could be bred by a factor of about twenty. For a preindustrialized nation, this was a tremendous economic accomplishment. Primarily funded by the Veterinary Department of the Russian Interior Ministry, Ivanov carried this technology to its next logical step, the creation of specialized hybrid animals for agricultural and industrial purposes, as well as for the sake of advancing the science. His artificial insemination experiments successfully crossed many closely related species: donkeys and zebras, mice and rats and other rodents, birds, and various species of cattle.

As early as 1910, Ivanov lectured on the possibility of crossing humans and apes, citing artificial insemination as the method of choice due to prevailing ethical objections to, well, interspecies partying, for lack of a better term. However, before he could make any progress, Ivanov's work came to an abrupt halt in 1917 with the Russian Revolution, which effectively dissolved most existing government programs and eliminated all of his funding. The new Soviet government was committed to technical innovation and science, but it took seven long years for Ivanov to rebuild his network of support. Ivanov's entire career could be fairly characterized as a constant fundraising effort, desperately seeking resources for his hybridization dream and other projects, and failing nine times out of ten. He should have been so lucky as to have the government come to him with an offer, much less an order.

Interestingly, many modern articles about Ivanov portray his work as a religiously motivated crusade. It's often said that the Russian and Soviet governments funded Ivanov not for any practical purpose, but merely out of atheist activism to prove evolutionary biology and to show that creationism has no place. Amid the developing nation's immense problems with famine and agricultural development, this would seem to be a bizarre reason to explore the capabilities of animal insemination. Nevertheless, there's an element of truth to it. In voicing support for Ivanov's 1924 grant proposal, the representative of the Commissariat of Agriculture said:

"...The topic proposed by Professor Ivanov...should become a decisive blow to the religious teachings, and may be aptly used in our propaganda and in our struggle for the liberation of working people from the power of the Church."

It's not clear whether this was the Commissariat's actual position or whether it was simply a sales tactic; either is plausible. Ivanov himself is not known to have ever expressed interest in this interpretation of his work; after all, he'd been studying reproduction as a scientist for almost 30 years, since long before the Soviet state existed.

It took another year for this particular proposal to be funded. Apes were prohibitively expensive and rare in Russia, so Ivanov set off for Africa to set up a new lab. After some false starts, he finally launched his own facility in Guinea with chimpanzees netted for him by local hunters. Using sperm from an unidentified man, Ivanov made three artificial insemination attempts on his female chimps. Because Ivanov observed that the local Africans viewed chimps as inferior humans, and viewed humans who had had contact with chimps as tainted, he performed these inseminations in secret with only his son present as an assistant. Ivanov knew that a mere three attempts was inadequate to hope for any success, but the difficulties and expenses of maintaining and inseminating the chimps was too great. So he conceived a more sustainable experimental technique: Collecting the sperm of only two or three male apes, and then using that to artificially inseminate human women.

He found no support for his plan in Africa — in large part because he had proposed to inseminate women in hospitals without their knowledge or consent — so he returned to the Soviet Union with his remaining chimps and founded a primate station in Sukhum (today called Sukhumi) on the Black Sea. Only one mature male survived, an orangutan named Tarzan. By 1929, the plan was to have five women be artificially inseminated, and then live at Ivanov's institute with a gynecologist for one full year. But just as the first woman volunteer was secured, known only to history as "G", Tarzan died. Ivanov ordered five male chimps, but just as they were delivered, his life suddenly turned in a new direction, driven by the constant turmoil of philosophies and favoritisms in the Soviet Union. Ivanov was accused of sabotaging the Soviet agricultural system and various political crimes, leading to his arrest a few months later. G never visited the Sukhum station, and no sperm was ever harvested from the new chimps. Ivanov died after two years of exile.

Ivanov's primate station survived, however, and became his only real legacy. By the 1960's it had over two thousand apes and monkeys, and was employed by the Soviet and American space programs. But nobody ever followed his ape-human hybrid research there, though conventional artificial insemination was often employed among its primate population.

So, does this history support or contradict the claim that Stalin wanted an ape-man hybrid race of slave super warriors? Well it certainly doesn't confirm it. Contrary to the modern version of the story, Stalin personally had no connection with Ivanov or his work, and probably didn't even know about it. No evidence has ever surfaced that Stalin or the Soviet government ever went out looking for someone to create an ape-man super soldier, though it's certainly possible that someone evaluating Ivanov's proposal may have made such an extrapolation.

Yet, in 2005, the Scottish newspaper *The Scotsman* reported the following:

> The Soviet dictator Josef Stalin ordered the creation of Planet of the Apes-style warriors by crossing humans with apes, according to recently uncovered secret documents. Moscow archives show that in the mid-1920s Russia's top animal breeding scientist, Ilya Ivanov, was ordered to turn his skills from horse and animal work to the quest for a super-warrior.

The latter claim, that Ivanov was "ordered" to shift his work, we've found to be demonstrably untrue. The former claim, that "secret documents" have been uncovered in Moscow, is a little hard to swallow. The article gives no information whatsoever about these alleged documents, and no source is even mentioned. A search of Russian language newspapers reveals no news stories about this at all, prior to *The Scotsman's* article. Certainly there are documents somewhere pertaining to the grants Ivanov received from both the Russian and the Soviet governments, but if these are what *The Scotsman* referred to, they are wrong when they describe them as secret, as recently uncovered, and that they showed Ivanov was ordered to create a super-warrior.

From what I can see, *The Scotsman's* story was merely another in a long line of cases where a journalist fills a slow news day with a sensationalized and/or fictionalized version of very old news, just as the *National Enquirer* did with the Roswell UFO story in 1978. In that case, the TV show *Unsolved Mysteries* picked it up and broadcast an imaginative reconstruction based on the article, and launched a famous legend. In this instance, the show *Monster Quest* picked up *The Scotsman* article and broadcast a 2008 episode called *Stalin's Ape Man*. The Internet has been full of articles about Stalin's supposed experiments ever since. Interestingly, a very thorough and well researched episode of *Unsolved History* on the Discovery Channel called *Humanzee*, which was all about human-ape hybrid experiments, did not mention Stalin or the Ivanov experiments at all. Why not? Because it was made in 1998, seven years before *The Scotsman* published its unsourced article, and introduced a new fiction into pop culture.

Oliver

Humanzee focused on a particular chimp named Oliver, still living as of today, who has a bald head, prefers to walk upright, and has a number of other eerily humanlike tendencies. Although Oliver has been long promoted as a hybrid, genetic testing found that he is simply a normal chimp. This result was disappointing to cryptozoologists and conspiracy theorists, but it did not surprise primatologists who knew that each of Oliver's unusual features is within the range of normal chimps. In fact, this was established 20 years ago by testing done in Japan, and again in 1996; it's just that nobody reported it since it was not the sensational version of the story.

Oliver is not a hybrid; Ivanov produced no hybrids; and other scientists have at least looked into it and never created any. There are the usual unsourced stories out of China of hybrids being created in labs, and even one from Florida in the 1920's. Desmond Morris, author of *The Naked Ape*, reported rumors of unidentified researchers in Africa growing hybrids, but even he dismissed it as "no more than the last quasi-scientific twitchings of the dying mythology." None of these tall tales are supported by any meaningful evidence.

But is it possible? Biologists who have studied the question are split, but the majority appear to think it is not, at least not from simple artificial insemination. But one conclusion can be drawn as a certainty, at least to my satisfaction: The urban legend that Stalin ordered Ivanov (or anyone else) to create an ape-man super soldier is patently false. It has all the hallmarks and appearance of imaginative writers creating their own news, and it was done at the expense of Il'ya Ivanov, whose proper place as a giant in the field of biology has been unfairly overshadowed by a made-up fiction. Treat this one as you would any urban myth: Be skeptical.

References & Further Reading

Davis, W. "Hybridization of Man and Ape to Be Attempted in Africa." *Daily Science News Bulletin*. 1 Jan. 1925, Number 248: 1-2.

Hall, L. "The Story of Oliver." *Primarily Primates Videos*. Primarily Primates, 21 Jan. 2008. Web. 15 Aug. 2010. <http://www.primarilyprimates.org/videos/ppvid_Oliver.htm>

MacCormack, J. "Genetic testing show he's a chimp, not a human hybrid." *San Antonio Express-News*. 26 Jan. 1997, Newspaper.

Morris, Desmond and Ramona. *Men and Apes*. London: Hutchinson, 1966. 82.

Rossiianov, K. "Beyond species: Il'ya Ivanov and his experiments on cross-breeding humans and anthropoid apes." *Science in Context*. 1 Jun. 2002, Volume 15, Number 2: 277-316.

Schultz, A. "The Rise of Primatology in the Twentieth Century." *Proceedings of the Third International Congress of Primatology, Zurich*. 1 Jan. 1970, Volume 2, Number 15.

Stephen, C., Hall, A. "Stalin's half-man, half-ape super-warriors." *The Scotsman*. 20 Dec. 2005, Newspaper.

34. Yonaguni Monument: The Japanese Atlantis

A look at a massive stone structure off the coast of Japan, said to be a manmade pyramid.

About 25 meters beneath the waters off Japan lies a stepped pyramid. We don't know who built it, or when; but there it is, plain as day, available for anyone to go down and inspect. Even now at this very minute, the current washes past sharply squared stone blocks standing dark and forbidding, rising nearly high enough to break the surface. It is called the Yonaguni Monument.

The Japanese archipelago stretches for nearly 4,000 kilometers, from Russia's Kamchatka Peninsula to the island of Taiwan, off the coast of mainland China. At its extreme southwestern tip is the small island of Yonaguni, Japan's most western point, just a scant 100 kilometers from Taiwan. It's quite small, less than thirty square kilometers, with only 1700 residents, but it's famous for something found in its waters: Hammerhead sharks.

Lots of hammerhead sharks. They're so ubiquitous that divers come from all over the world to swim with them. And wherever you have a lot of divers, things under the water tend to be found. And that's just what happened in 1986, when a representative from the Yonaguni tourism board was out exploring off the southernmost tip of the island, looking for a hammerhead diving spot to promote. What he came across was not what he set out to look for, though.

As you're probably aware, Japan is in a region of great tectonic instability, the Pacific Ring of Fire. It lies just beside the convergence of the Pacific Plate and the Philippine Sea Plate, and as a result, it's home to ten percent of the world's active volcanoes. Severe earthquakes are a familiar event there. The layered sandstone bedrock around Yonaguni is therefore deeply fractured. As the tourism rep swam, he passed over this cracked and piled terrain, until he came to a particular formation that stood out. He named the area Iseki Point, or Ruins Point.

He passed the word that he'd found something that looked like a manmade castle. A professor of marine geology, Masaaki Kimura, came to have a look for himself, and what he saw has dominated his life ever since. Kimura founded the Marine Science and Cultural Heritage Research Association, an organization devoted to proving that the Yonaguni Monument is not merely the natural formation it would appear to be, but rather a manmade structure, consisting of a huge network of buildings, castles, monuments, a stadium, and other structures, all connected by an elaborate system of roads and waterways.

A diver examines Yonaguni

It's exactly the kind of story that the public loves. Headlines trumpeted Kimura's discovery with such cliché phrases as "Scholars mystified", "underwater city", and "Japanese Atlantis" (as I so cleverly titled this chapter). *History's Mysteries* on the History Channel produced an episode called "Japan's Mysterious Pyramids" which promoted the idea with little critique; and again on *Ancient Discoveries* with an episode called "Lost Cities of the Deep". The BBC and the Discovery Channel have also produced documentaries promoting the Yonaguni Monument's manmade past.

Web forums and conspiracy sites love to exaggerate such stories as this one. Among the formations identified by Dr. Kimura is one that he has named "Jacques' Eyes", after Jacques Mayol who used to freedive the site. It's a big roundish rock with two depressions near where eyes might be, but it certainly does not look like a carved head and Kimura does not presume to identify it as one. He has a photograph of it on his web site that he took personally. He contends that the eyes were carved, but that the rest of the rock is natural. However, there's a completely different photograph floating around the Internet showing three divers swimming around a tremendous stone head that is very obviously manmade, including what looks to be a feathered headdress. Whatever the source of this photograph is, it bears no resemblance at all to the rock at Yonaguni, despite its being so identified on every web site I found it.

I've studied Dr. Kimura's photograph of the Jacques' Eyes formation and I'm far from convinced the eyes were carved. They're large concave depressions without distinct edges, not eye shaped, not symmetrical, and not convex like an eyeball. I believe that even incompetent artists would have done a far better job of representing human eyes. Although the underwater lighting is from directly above and the shadows can make them resemble eyes, they wouldn't have looked anything like that in the open sunlight.

And open sunlight is the key to Dr. Kimura's hypothesis, which is that this formation was on dry land when ocean levels were lower during the last ice age. 8-10,000 years ago, the Yonaguni Monument was dry; and for tens of thousands of years before that, it was high and dry.

As you can guess, I'm not the only one who is skeptical of Dr. Kimura's interpretation of the bedrock formations. Virtually all marine geologists who have seen the pictures are satisfied that it's perfectly consistent with other formations of fractured sandstones. Everyone grants that it is unusually dramatic and has a lot of interesting features, but there's nothing there that's not seen anywhere else. The work of Kimura's own foundation, which researches many similar formations off the surrounding islands, is evidence that Yonaguni is not especially unique.

This dispute plays right into the hands of the documentary filmmakers, who are looking for the conflict angle in order to promote the idea of controversy, trying to convince us that scientists are somehow torn or debating over this. They're not. Kimura has a few supporters, but the consensus is resoundingly against him. Dr. Robert Schoch, a geologist at Boston University, is the most often quoted scientist taking the opposing position. Dr. Schoch is probably best known for his work on assigning Egypt's Sphinx and Great Pyramid dates that are much earlier than previously believed, based on his analysis of weathering (you may have seen him discuss this on science channel documentaries). So Schoch is, himself, a bit of a maverick; apparently very few other geologists or archaeologists have found Kimura's photographs and interpretations to be compelling enough to work on.

Schoch has made a few dives on Yonaguni; Kimura has made over a hundred. Nevertheless, Schoch noted what is, I think, the single most damning point against the idea that Yonaguni is manmade:

> "...The structure is, as far as I could determine, composed entirely of solid 'living' bedrock. No part of the monument is constructed of separate blocks of rock that have been placed into position. This is an important point, for carved and arranged rock blocks would definitively indicate a man-made origin for the structure - yet I could find no such evidence."

The paleogeology of the region is well known, and Schoch brought samples of the Yonaguni rock to the surface for analysis. He found that they were, as suspected, mudstone and sandstone of the formation called the Lower Miocene Yaeyama Group, which was deposited some 20 million years ago.

> "These rocks contain numerous well-defined, parallel bedding planes along which the layers easily separate. The rocks of this group are also criss-crossed by numerous sets of parallel and vertical ... joints and fractures. Yonaguni lies in an earthquake-prone region; such earthquakes tend to fracture the rocks in a regular manner... The more I compared the natural, but highly regular, weathering and erosional features observed on the modern coast of the island with the structural characteristics of the Yonaguni Monument, the more I became convinced that the Yonaguni Monument is primarily the result of natural geological and geomorphological processes at work. On the surface I also found depressions and cavities forming naturally that look exactly like the supposed 'post holes' that some researchers have noticed on the underwater Yonaguni Monument."

In recent years, Dr. Kimura has acknowledged that the basic structure of the Monument is probably natural, but asserts that it has been "terraformed" by humans, thus creating the specific details such as Jacques' Eyes and the roads. He has also found and identified what he believes to be quarry marks and writing. To my eye, these don't look anything like quarry marks or writing. It's not a testable claim; the analysis simply comes down to personal opinion and interpretation. But it's certainly possible. Were there people living there 8-10,000 years ago?

From everything we know so far, the answer is no. Yonaguni is one of the Ryukyu Islands, of which Okinawa is the largest, and the earliest archaeological evidence is that of the Late Shellmound phase which began only as recently as 300 BCE. The Ryukyu Islands are in deep water, at least 500m deep on all sides, and at no time during the last glacial age were the islands accessible by land bridge. This means that if any people were there when Yonaguni was on dry land, they did not stay, and they would have to have arrived by boat. This is something else we can check.

Nearby Taiwan has probably been populated since Paleolithic times, tens of thousands of years ago, but the earliest population for which we have any evidence was the Dapendeng

Culture which began 7,000 years ago. This is about the time that fishermen began to use canoes for coastal travel, about 5000 BCE. If the Dapendeng colonized Yonaguni, they would have had to have done so by boat. This cuts the timing very, very close. Yonaguni was probably already awash when the first Dapendeng canoes put to sea as glacial melt brought sea levels up. Of course, the studies which give us those dates could be wrong. But we do know that if the Dapendeng ever did colonize Yonaguni or the Ryukyus, they did not stay. Genetic studies have shown that the founding Ryukyu populations migrated southward from Japan, not from Taiwan.

So taking everything into account, the likelihood that prehistoric human hands ever had the opportunity to touch the stones of the Yonaguni Monument appears vanishingly small. The only evidence that they did is personal assessment of some fairly ambiguous undersea formations, none of which are geologically surprising, and all of which have analogs at known natural sites around the world. If the Yonaguni Monument is truly a Japanese Atlantis, it is a highly improbable one indeed.

REFERENCES & FURTHER READING

Chang, K. "The Formosa Strait in the Neolithic Period." *Kaogu.* 1 Jun. 1989, Number 6: 541-550, 569.

Hudson, M., Takamiya, H. "Dental pathology and subsistence change in late prehistoric Okinawa." *Bulletin of the Indo-Pacific Prehistory Association.* 1 Jan. 2001, Volume 21: 68-76.

Jiao, T. *Lost Maritime Cultures: China and the Pacific.* Honolulu: Bishop Museum Press, 2007.

Kimura, M. "Yonaguni." *Marine Science and Cultural Heritage Research Association.* Dr. Masaaki Kimura, 24 Oct. 2007. Web. 20 Aug. 2010. <http://web.mac.com/kimura65/>

Milne, G., Long, A., Bassett, S. "Modelling Holocene relative sea-level observations from the Caribbean and South America." *Quaternary Science Reviews.* 1 Jan. 2005, Volume 24, Numbers 10-11: 1183-1202.

Schoch, R. "An Enigmatic Ancient Underwater Structure off the Coast of Yonaguni Island, Japan." *Circular Times.* Dr. Colette M. Dowell, 19 Apr. 2006. Web. 20 Aug. 2010. <http://www.robertschoch.net/Enigmatic%20Yonaguni%20Underwater%20RMS%20CT.htm>

35. The Myers-Briggs Personality Test

A critical look at the world's most popular psychological metric, the Myers-Briggs Type Indicator.

In this chapter, we're going to delve into the murky depths of Jungian psychology, and examine one of its most popular surviving manifestations. The Myers-Briggs test is used all over the world, and is the single most popular psychometric system, with the full formal version of the test given more than 2,000,000 times a year. But is it a valid psychological tool, is it just another pop gimmick like astrology, or is the truth somewhere in between?

The Myers-Briggs Type Indicator, called MBTI for short, more properly owes the bulk of its credit to the great Swiss analytical psychologist Carl Jung. In 1921, Jung published his book *Psychological Types*, in which he laid out all the same concepts found in the MBTI, but he had them organized quite differently. Jung had everyone categorized as either a "perceiver" or a" judger". Perceivers fell into one of two groups: sensation and intuition; while judgers also fall into two groups: thinking and feeling. So everyone fits into one of those four buckets. Finally, each bucket is divided into two attitude types: introversion and extraversion. Thus, the scale proposed by Jung divided us all into one of eight basic psychological types.

An American woman, Katherine Briggs, bought Jung's book and was fascinated by it. She recommended it to her married daughter, Isabel Briggs Myers, who had a degree in political science. The two of them got hooked on the idea of psychological metrics. Together they sat down and codified their own interpretation of Carl Jung, making a few important changes of their own. Jung had everyone fitting into one of four basic buckets. Myers and Briggs decided that each person probably combined elements, so they modified Jung's system and made it a little more complex, ending up with four dichotomies, like binary switches. Any combination of the four switches is allowed, and Myers and Briggs reasoned that just about every personality type could be well described by one of the sixteen possible ways for those switches to be set. Basically, according to Myers and Briggs, we're all represented by a four-digit binary number.

- The first dichotomy is called your *Attitude*, and according to the MBTI, you're either an E for Extravert or an I for Introvert. Extraverts prefer action, frequent interaction, focus outward, and are most relaxed when interacting with others. Introverts prefer thought, less frequent but more substantial interaction, and are most relaxed spending time alone.

- The second dichotomy is your *Perceiving* function, and you're either S for Sensing or N for Intuition. Sensing is the scientific, tangible data-driven approach to gathering information, preferring to deal in concrete, measurable information. The Intuition approach prefers theoretical, abstract, hunch-driven information, finding more meaning in apparent patterns and context.

- The third dichotomy is your *Judging* function, and you're either a T for Thinking or an F for Feeling. This is basically how you make decisions. Thinking makes the logical decision, what's best for the situation, based on rules and pragmatism. Feeling decides based on empathy for the people whom the decision affects, seeking balance and harmony.

- The fourth and final dichotomy is your *Lifestyle*, and you're either a J for Judgment or a P for Perception. This one gets a little confusing. Judgment types prefer to use the third dichotomy, *Judging*, when relating to the outside world, while Perception types prefer the second *Perceiving* dichotomy; but how that preference is determined is based on whether you're an Introvert or an Extravert. Suffice it to say, for the purpose of this light overview, that this last of the four dichotomies, *Lifestyle*, is the most complicated; and it's where Myers and Briggs most creatively expanded upon Jung on their own.

The basic test, of which there are several variations and revisions, is called the MBTI Step I and it's a series of almost 100 questions, each with two possible answers. Each question consists of two short statements or word choices, and you simply choose which of the two you prefer. When the results are tabulated, you should ideally have your preference established for each of the four dichotomies; and congratulations, you are now identified by one of sixteen possible personality types. Myers and Briggs gave names and descriptions to all sixteen, names such as the Executive, the Caregiver, the Scientist, and the Idealist.

Perhaps the most common misconception about the MBTI is that it shows your aptitude, helping you determine what kinds of things you'd be good at. This is not the case. Myers-Briggs is only about determining your preference, not your ability. There might be things that you're good at that you don't enjoy, and there might be things you enjoy that you're not good at. The MBTI helps your find your comfort zone, the types of activities you'll like and be most content with; not necessarily those at which you'll be especially competent.

Even though neither had any background in psychology, Myers and Briggs enjoyed great success with their system. As Mrs. Briggs was getting quite old, Isabel Myers was the main driving force. Her initial idea was that certain personality types would more easily excel at different jobs, and the tool was intended to be used by women entering the workforce during World War II. However, it was not published until 1962, but since that time, it's become the most widely used basic psychology test. It's most often used outside of the psychological profession, and is employed in career counseling, sports coaching, marriage counseling, dating, professional development, and almost every other field where people hope to be fit with a role that would work best for them.

So the MBTI's practical use is overwhelmingly unscientific, and it's often criticized for this. Criticism ranges from the pragmatic fact that neither Jung nor Myers and Briggs ever employed scientific studies to develop or test these concepts, relying instead on their own observations, anecdotes, and intuitions; all the way to charges that your MBTI score is hardly more meaningful than your zodiac sign.

One obvious trait that the MBTI has in common with horoscopes is its tendency to describe each personality type using only positive words. Horoscopes are so popular, in part, because they virtually always tell people just what they want to hear, using phrases that most people generally like to believe are true, like "You have a lot of unused potential." They're also popular because they are presented as being personalized based on the person's sign. This has been called the Forer Effect, after psychologist Bertram Forer who, in 1948, gave a personality test to his students and then gave each one a supposedly personalized analysis. The impressed students gave the analyses an average accuracy rating of 85%, and only then did Forer reveal that each had received an identical, generic report. Belief that a report is customized for us tends to improve our perception of the report's accuracy.

I notice this right away when I read Isabel Myers' description for my own personality type, ISTJ, the *Duty Fulfiller:* "Practical, matter-of-fact, realistic, and responsible." Basically it's a nice way to say "Dry, boring, and punctual," which hits my nail pretty squarely on the head. From that alone, I might conclude that the MBTI is extraordinarily insightful. But if I look at her description of my opposite counterpart, an ENFP, the *Inspirer*, that person is "Warmly enthusiastic and imaginative. Sees life as full of possibilities." Who wouldn't like to believe that about his or her self? If I'd taken the test and been handed that result, I might be equally inclined to embrace it, probably thinking something like "Wow, I'm even more awesome than I thought I was."

Due to these legitimate criticisms of the MBTI and its unscientific underpinnings, the test is rarely used in clinical psychology. I did a literature search on PubMed and discovered that, interestingly, many of the published studies of its practical utility come from nursing journals. Many of the other publications pertain to relationship counseling and religious counseling. Normally, this is a red flag. When you see a topic that purports to be psychological being used in practically every professional discipline *except* psychology, you have very good reason to be skeptical of its actual value. Should we dismiss the Myers-Briggs Type Indicator as a psychometric?

The test does have some severe inherent problems. It's been found that 50% of test takers who retake it score differently the second time. This is because nobody is strictly an E or an I, for example, but somewhere in between. Many people are right on the border for some of the four dichotomies, and depending on their mood that day or other factors, may answer enough questions differently to push them over. Yet the results inaccurately pigeonhole them all the way over to one side or the other. This makes it possible for two people who are very similar to actually end up with completely opposite scores. Isabel Myers was aware of this limitation, and did her best to eliminate questions that did not push people away from the center when the results were studied in aggregate. It was a hack.

From the perspective of statistical analysis, the MBTI's fundamental premise is flawed. According to Myers and Briggs, each person is either an introvert or an extravert. Within each group we would expect to see a bell curve showing the distribution of extraversion within the extraverts group, and introversion within the introverts. If the MBTI approach is valid, we should expect to see two separate bell curves along the introversion/extraversion spectrum, making it valid for Myers and Briggs to decide there are two groups into which people fit. But data have shown that people do not clump into two separately identifiable curves; they clump into a single bell curve, with extreme introverts and extreme extraverts forming the long tails of the curve, and most people gathered somewhere in the middle. Jung himself said "There is no such thing as a pure extravert or a pure introvert. Such a man would be in the lunatic asylum." This does not support the MBTI assumption that people naturally separate into two groups. MBTI takes a knife and cuts the bell curve right down the center, through the meatiest part, and right through most people's horizontal error bars. Moreover, this forced error is compounded four times, with each of the four dichotomies. This statistical fumble helps to explain why so many people score differently when retaking the test: There is no truly correct score for most people, and no perfect fit for anyone.

And this has been borne out in observation. A number of studies have found that personality types said to be most appropriate for certain professions, notably nursing or teaching, turn out to be no more prevalent among that profession than among the general population. The Army Research Institute commissioned one such study to determine if the MBTI or similar tests could be used to improve the placement of personnel in different duties, and firmly concluded that the results of such tests did not justify their use in career counseling.

From reviewing the literature, I do find one common theme among mainstream psychotherapists where the use of the MBTI is advised, and that's as a conversation starter. It's a fine way to give people a quick snapshot of what their strengths and weaknesses *might* be, and of those with whom they interact. To get the dialog going, this is a perfectly valid tool. But as a tool for making career decisions, relationship decisions, or psychiatric assessment, no. Although it would be nice to have a magically easy self-analysis tool that can make your decisions for you and be your crystal ball, the Myers-Briggs test is not it. It is interesting and it does have value as a starting point for meaningful dialog, but that's where the line should be drawn.

REFERENCES & FURTHER READING

Dickson, D., Kelly, I. "'The Barnum Effect' in Personality Assessment: A Review of the Literature." *Psychological Reports.* 1 Feb. 1985, Volume 57, Number 2: 367-382.

Druckman, D., Bjork, R. *In the Mind's Eye: Enhancing Human Performance.* Washington, DC: National Academy Press, 1991.

Howes, R., Carskadon, T. "Test-Retest Reliabilities of the Myers-Briggs Type Indicator as a Function of Mood Changes." *Research in Psychological Type.* 1 Jan. 1979, Volume 2, Number 1: 67-72.

Jung, C. *Psychological Types.* New York: Harcourt, Brace & Company, Inc., 1923.

Long, T. "Myers-Briggs and Other Modern Astrologies." *Theology Today.* 1 Oct. 1992, Volume 49, Number 3: 291-295.

Myers, Isabel and Peter. *Gifts Differing.* Palo Alto: Consulting Psychologists Press, 1980.

36. Toil and Trouble: The Curse of Macbeth

The Curse of Macbeth should make Shakespeare's play too dangerous to perform. Is this the case?

They simply call it "The Scottish Play", because even to utter the title of Shakespeare's *Macbeth* is to invite bad luck. The very same bad luck, in fact, that has plagued performances throughout its history, according to theater lore. From tragedies onstage to deaths and riots surrounding performances, the curse of *Macbeth* is one of the most enduring superstitions of the stage, and seems to be taken quite seriously.

The basic claim is that performing *Macbeth*, or even speaking its title in a theater, invokes an ancient curse as old as the play. This curse strikes actors or other people associated with the performance, sometimes killing or maiming them. This curse, so goes the tale, has its roots in the play's occult storyline of witchcraft, murder, and ghosts. The most often cited reason for the curse is a belief at the time that Shakespeare had used real witches' incantations in the famous scene where the three witches chant:

Double, double toil and trouble; Fire burn, and cauldron bubble.

The legend is that Shakespeare wanted to throw something special into the play to please King James, who had written the 1597 book *Daemonologie* which discussed witchcraft and warned against its use. So Shakespeare used some of King James' documented incantations in the scene, probably hoping to ingratiate himself with the King. No good evidence exists for this, but the story maintains that some practicing witches saw the play and took great offense at this misuse of their sacred craft, and placed a curse upon any who might perform *Macbeth*. Now, whenever the play is given, the three witches whose spells were appropriated are awoken and it is they who cause the disasters onstage.

If you search the Internet for examples of such disasters, you'll find plenty. I could rattle on for 25 minutes just listing some of the many that have been reported: Actors being killed or injured during the stage fights when real weapons were used by mistake, natural disasters happening during performances, accidents and illnesses striking the crew before and after shows; the list goes on and on until you're bored to death and would be glad to count yourself among the casualties. One such episode, however, deserves special mention for its extraordinarily high body count.

Often cited as the most dramatic evidence of the curse is a riot that erupted at the Astor Place Theater in New York in 1849. The National Guard was called and fired on the crowd, killing at least 25 people and injuring some 120, all due to rival support for two different actors playing *Macbeth* on the same night at two different theaters. At least, that's usually how it's framed by fans of the curse. In fact, the Astor Place Riot had everything to do with class struggles in New York City, and little to do with *Macbeth*. Unrest had been growing for years between the working class, which included many Irish immigrants, and the Anglophile upper class. The discontent was coming to a head, and the National Guard was already in place some days before the actual riot on *Macbeth's* opening night. Irish and American workers planned to express themselves by crashing the opening night of the upper class's favorite British actor, William Macready. They stoned and tried to burn down the theater, people started shooting guns, and by morning the cobblestones were awash with blood. The next night an angry mob demanded an explanation from the authorities, and more violence ensued, this time resulting in the death of a young boy.

The Astor Place Theater had been built, apparently, largely as a way for the well-heeled to have somewhere to go other than the Bowery Theater, traditionally the principal theater in town, but which catered to all classes. The rising American star with a blue collar image, Edwin Forrest, planned his opening on the same night as Macready's largely as a slap in the face to this rising elitist sentiment. So although the riots were technically touched off by performances of *Macbeth*, the play itself had nothing to do with them. The riots were due, and would have happened whether *Macbeth* existed or not. If the three witches had chosen this particular performance to cast spells and cause trouble, they would have been well advised to hide under a table.

Many people have tried to put forth rational explanations for the events attributed to the curse. Often cited is that Macbeth has a lot of dim lighting and fight scenes using stage weapons. Such weapons are still dangerous, just not very sharp; and you're bound to have statistically more injuries in any play that has weapon fights than in plays that don't. Statistically, we should also expect more falls and other onstage accidents in plays with dim lighting. Even in brightly lit plays, it's hard enough to see what you're doing onstage because of the stage lights shining in your eyes; in a dimly lit scene, you could easily be practically blind. I don't really buy the dim lighting explanation. Granted my own stage experience is fairly limited, but when I'm brightly lit is when I have the hardest time seeing. Dim lighting, even no lighting, lets my eyes adjust and I can see my way around much better than when I'm blinded by spotlights. Other stage performers' experiences may vary.

But let's stop here and think back to the skeptical process. One of our fundamental rules is that before trying to explain a strange event, you must first establish whether that strange event ever actually happened. In this case, we need not bother looking into the validity of the curse, or any other such thing, unless and until we've established that there is in fact a history of mysterious accidents associated with the performance of

Macbeth that deviates beyond the range of what typically happens in plays.

I was inspired by an earlier success I had when researching the curse of King Tut, when I discovered that a doctor had performed a retrospective cohort study on the people who were alleged to have fallen victim to that curse. He discovered that, when analyzed properly, the curse (if it existed) was not a terribly effective one. The lifespans of those who were exposed to the curse did not significantly differ from those who were not exposed. Encouraged by the publication of this study (it was in the *British Medical Journal*), I turned to all the scholarly sources to see if anyone had performed a proper statistical analysis of theater accidents, ideally involving *Macbeth*. I even assigned this task to my backup research team, a Google Groups list to whom I'll throw a question or two on occasion when I have trouble tracking something down. It's a heck of a list; hundreds of scientists and researchers in virtually every discipline, but even this mighty team came up short. We couldn't find any such research published anywhere. There is no end to scholarly articles discussing the curse: Lists of tragic events, Shakespeare's history with King James and the witches, and how to combat the curse (leave the theater, spin around, spout some profanity); but not a whisper inquiring into the proving curse's existence.

So, college students, there's a research project for you. This would not be easy. First you'd have to eliminate things like natural disasters that can't reasonably be attributed to the performance, and things like accidents striking people weeks after the play. To include these would require you to also correlate any other plays the victim may have attended, since it makes just as much sense to blame the accident on a different play he may have also seen in the same time period. Theaters may have records of accidents occurring on their premises, and those dates could certainly be matched with whatever was being performed. The number of cast and crew required would likely impact the chances of an accident for any given show, as would the use of dangerous equipment like lights, trap doors, flying harnesses, trickery scenery, and stage weapons. All of these things would need to be taken into account. It's little wonder that we couldn't find a record of such a study being performed.

Even the longest of the published lists of tragedies associated with *Macbeth* performances does not seem surprising, considering that we're talking about one of the world's most popular plays that has been performed constantly worldwide for more than 400 years. But since it appears that no analysis has been done, we can't conclude for certain that *Macbeth* is any more or less dangerous to perform than any other play.

The longest lists I've seen include perhaps twenty or thirty tragedies. Considering that there have been unknown tens of thousands (possibly hundreds of thousands) of *Macbeth* performances, the curse (if it exists at all) appears to be decidedly impotent. An actor probably stands a greater chance of having a car accident on his way to the theater than he has of being struck by the curse of *Macbeth* — that's another research project for a student.

Therefore, since we can't establish that the curse exists, we don't yet have a confirmed phenomenon to explain. We have a tall tale, told and retold over the centuries, and sufficient reason to suspect that wizened veteran actors may enjoy having a little fun at the expense of the newbies, perpetuating the story of a curse regardless of whether the actual number of associated accidents deviates from the norm.

Let's also keep in mind that the legend of the curse began almost right away when Shakespeare originally opened the play. Shakespeare was probably not a fool and knew that there's no such thing as bad publicity. What show's ticket sales would suffer if word got out that one of the actors might be accidentally killed during the performance? A little curse never hurt anybody (well, maybe a few).

So the curse of *Macbeth:* fact or fiction? I'm going to remain unconvinced that there's anything extraordinary going on, but will eagerly take a look at any good research that emerges. A list of anecdotes on the Internet is insufficient to prove that a supernatural force must be in effect.

References & Further Reading

Bernauw, P. "The Curse of Macbeth." *Unexplained Mysteries.* Unexplained-Mysteries.com, 10 Aug. 2009. Web. 4 Nov. 2010. <http://www.unexplained-mysteries.com/column.php?id=160421>

Cliff, N. *The Shakespeare Riots: Revenge, Drama, and Death in Nineteenth-Century America.* New York: Random House, 2007.

Crystal, D., Crystal, B. *The Shakespeare Miscellany.* Woodstock: Overlook Press, 2005. 176-177.

Homberger, E. *Mrs. Astor's New York: Money and Social Power in a Gilded Age.* New Haven: Yale University Press, 2004. 109-110.

Huggett, R. *Supernatural on Stage: Ghosts and Superstitions of the Theater.* New York: Taplinger, 1975.

Nelson, Mark R. "The Mummy's Curse: Historical Cohort Study." *British Medical Journal.* 21 Dec. 2002, 325(7378): 1482-1485.

37. The Frog in the Stone

Stories abound of living frogs being found encased in solid rock. What's really going on here?

In this chapter, we're going to take a rock hammer, split open a nodule of stone, and be amazed as a perfectly preserved frog hops out from a cavity within. This same surprising phenomenon has been reported since the fifteenth century, as many as 210 separate incidents, by some reports. If you believe the stories, frogs and toads have the ability to survive impossibly long time periods sealed inside solid rock. Examples include a toad extracted from a block of stone by excavators in Hartlepool, England in  1865; a British soldier working in a quarry during WWII who found one inside a slab of rock retrieved from 20 feet underground; and some French workmen who split open a flintstone in 1851 and found a living toad inside a cavity that perfectly matched its body contours. But these are only a few; many, many similar stories are found throughout the literature. With so many stories describing nearly the exact same phenomenon, one has to conclude that there must be something to it. Do frogs indeed have this astonishing ability?

A difficulty encountered when investigating this is that the majority of such stories are second and third hand reports from the 18th and 19th centuries. This means we don't have anything to examine, much less test. What we have is usually a pulp news reporter's interpretation of a story he heard from someone. That doesn't make it wrong, but it does add a chain of elements of unreliability to each story, and it certainly leaves us empty handed if we want to verify any part of it. We can, however, take a look at the science that *is* known. Is it possible for an amphibian — or any other animal — to actually go completely dormant for what seems an impossibly long time, and emerge alive?

There are two types of dormancy into which some animals enter to survive extreme conditions: hibernation, which happens during a cold winter; and estivation, a way of surviving a hot, dry summer. Both techniques involve a slowing of the metabolism, where breathing and the heartbeat slow down. The purpose of dormancy is to minimize the consumption of resources. These resources include both the animal's internal fuel stores and oxygen from the surrounding environment.

The purpose of hibernation is to escape the need to burn too much fuel to stay alive during a cold spell when adequate food is not available. Many frogs hibernate underwater. They slow their metabolism enough that they can get the oxygen they need through their skin directly from the water itself. Some will cover themselves partially with mud, but not entirely, because they need exposure to the oxygenated water. Some can also hibernate on land, usually under leaves or other woodland debris where they're still exposed to oxygen. They'll usually continue moving about, though slowly. No matter how cold it gets, most species won't freeze solid because of high glucose levels that lowers their bodies' freezing points, like antifreeze in your car. If a pond freezes over completely and the oxygen in the water is depleted, a hibernating frog will die. How long they hibernate depends on the temperature and region, but about half a year is the maximum. Longer than that, the frog runs out of calories and starves to death.

During the hot summer, a few species of frogs and toads estivate, which is all about preserving the body's water. They start with large water reserves in their bladder, then burrow down into the soil where it's cooler and more humid. Some species shed their outer layers of skin. These layers become a sort of watertight cocoon that seals the animal's hydration inside, all except for its nostrils which it needs to breathe. When the rains return, the animal breaks out of its cocoon and returns to the surface. Since estivating frogs and toads keep their nostrils open, oxygen is not a problem, and they can remain in this state as long as their calories and water hold out. Their metabolism drops to about 20-30% of normal. Estivation can last a maximum of about 10 months before stores run out, with the animal losing 50% of its total body mass.

Hibernation and estivation are the two known states into which frogs and toads can go to minimize their energy expenditure for extended periods. But if we want to explain the anecdotes telling of these animals emerging alive after decades, centuries, or even millennia of entombment in rock, we have to look elsewhere. Is it possible for a frog or toad to be completely hermetically sealed and preserved alive? There would be no water loss because there's nowhere for the water to go; but there's also no oxygen. Can the animal shut off completely, and remain alive?

In both hibernation and estivation, lipids are the primary fuel source, with fatty acid oxidation being the primary use of the energy budget. Proteins from muscle tissue break down into the amino acid glutamate, which is also oxidized as a secondary fuel source. Along with a few other less significant energy sources, these chemical reactions keep the organs alive during the torpid period, but they're also needed to fuel the animal's arousal when it wakes up. The "frogs in stones" story requires these processes to stop completely, and to burn zero resources indefinitely, and then restart once the animal is exposed to oxygen as the rock is broken open.

Here's the problem with that. If all metabolic activity goes to zero, that's basically the definition of death, and the body is then chemically and biologically indistinguishable from a corpse. The various components of the immune system would stop as well. With zero metabolism, cells enter autolysis, in which the cell's own enzymes break down the cell tissue itself. Soon the only activity that would remain would be bacteriological. Like all animals, amphibians have bacteria throughout their systems. Bacteria serve crucial functions such as fighting infections and aiding in digestion. When an animal dies, one of the first things that happens is that the anaerobic bacteria, which do not require oxygen, break down the body's tissues. In the hermetically sealed environment, the bacteria would eventually die as well, but only after consuming all available resources; in this case, the body of the frog. To put it succinctly, if your metabolism stops, you will decompose, and you cannot live through it.

But let's give the benefit of the doubt to the frog, and assume there is some way — as yet unknown to science — that frog biology can avoid both autolysis and bacterial destruction when all metabolism has ceased. Let's even take a biologically implausible leap and grant that they've stayed hydrated and oxygenated, and have not consumed their stores of lipids and proteins. The stories tell of the frogs reanimating, even springing out of their stone tombs and hopping around. This energy exertion requires metabolism, and it requires that the muscles be biochemically ready for action. The only way this can be is if oxidation had been taking place for some time. In other words, the frog would have required substantial advance notification that he was about to be released from the rock, and his metabolism would have had to have found some way to jump start itself. It's hard to trigger something when you're biologically dead and have zero energy with which to do it.

My opinion is that the stories of frogs and toads entombed within solid rock are false, at least the way they've been reported. Some are known hoaxes, notably one example taken on tour by Charles Dawson (who also brought us the Piltdown Man hoax) and pictured all over the Internet in credulous articles. Fabrications aside, there are probably other explanations for what the witnesses perceived. Perhaps the material in which the animal was found was laid very recently, during the current construction project. Maybe it was not a hole but the end of a tunnel, the rest of which was destroyed by the excavation. Perhaps the frog came from elsewhere, and given the simultaneous timing of the witness splitting open the rock and noticing the frog, the witness erroneously concluded the frog must have come from the rock. We can speculate all day long, but there's minimal value in doing so. The fact is that all such stories — those that were not fabrications — were in different circumstances and probably had different explanations. I strongly doubt that a single phenomenon was the cause of every story, and my doubt is even stronger that the least plausible possibility is the one that actually happened.

The value of anecdotal stories like these is not as evidence, but to suggest directions for future research. If frogs could be proven to have this ability, it would overturn much of what we've learned, and would be a tremendous discovery for a biologist. If one is cold-hearted enough to experiment, it should be a simple matter to encase various species of frogs and toads into various types of stone-like materials, expose them to various conditions, and then crack them open in two years, five years, twenty years. Comparative tissue testing could establish whether they were able to get their metabolisms to stop completely, and then restart. If this proves to be the case, I will enthusiastically change my opinion.

And it turns out that at least one researcher has done just that. In 1825, the Rev. Dr. William Buckland, the English paleontologist who first described a fossilized dinosaur, buried 24 toads, each in its own block of either sandstone or limestone, each sealed with a glass window. He checked them one year later. All those buried in sandstone were dead and decomposed, while a few of those entombed in the oxygen-permeable limestone had successfully estivated and were still alive, though badly emaciated. He reburied the survivors for another year, and found them all dead. That sounds about like what we'd expect, given our knowledge of estivation. Stories tell of similar experiments done by others including the English naturalist Edward Jesse, often with better results, and all with (unfortunately) dubious credibility. Experimentation has — so far — failed to replicate what people think is happening, suggesting that the explanation for these stories may lie outside of the actual entombment of living frogs.

It's often been pointed out that astronomers, who spend more time looking at the sky than the average person, also report far fewer UFOs, because they actually know what they're looking at. We see a similar effect with entombed frogs. In all of the many reports, not one comes from a trained paleontologist or field geologist, who spend far more time carefully chipping away at rocks and paying close attention to what they're doing. That doesn't prove anything, but it does remind us that when a story comes from everyone except the experts, you generally have very good reason to be skeptical.

References & Further Reading

Bondeson, J. "The Toad in the Hole." *The Fortean Times*. Dennis Publishing Limited, 1 Jun. 2007. Web. 10 Sep. 2010. <http://www.forteantimes.com/features/articles/477/toad_in_the_hole.html>

Buckland, F. "Curiosities of Natural History." *Blackwood's Edinburgh Magazine*. 31 Dec. 1858, Volume 83: 306-361.

Emmer, R. "How do frogs survive winter? Why don't they freeze to death?" *Scientific American*. Nature America, Inc., 24 Nov. 1997. Web. 6 Sep. 2010. <http://www.scientificamerican.com/article.cfm?id=how-do-frogs-survive-wint>

Sheaffer, R. "Psychic Vibrations." *Skeptical Inquirer*. 1 Jun. 1986, Volume 10, Number 4: 308.

Storey, K. "Turning Down the Fires of Life: Metabolic Regulation of Hibernation and Estivation." *Molecular Mechanisms of Metabolic Arrest*. 1 Jan. 2000, BIOS Scientific Publishers: 1-21.

Wagner, S. "Animals Sealed in Stone." *About.com*. The New York Times Company, 18 Sep. 2005. Web. 8 Sep. 2010. <http://paranormal.about.com/od/earthmysteries/a/aa011704.htm>

38. The Brown Mountain Lights

A ghost light in North Carolina has people scratching their heads. What's the explanation?

Come with us now to the Appalachian Mountains in North Carolina, about 110 kilometers northwest of Charlotte. Within the Pisgah National Forest are rocky gorges, streams, and green everywhere you look. Hikers and backpackers abound. Some of the lucky night time visitors — or perhaps the unlucky, depending on your perspective — may get more than they bargain for. For the region is home to one of the world's infamous ghost lights, the Brown Mountain Lights.

We've covered "ghost lights" on Skeptoid before; specifically the Marfa Lights in Texas and the Min Min Light in Australia. In both cases, those lights were conclusively found to be superior mirages of actual lights below the horizon. The terrain in those places are similar; both are deeply cleft with gulleys which trap hot air from the day which is then overlaid with cold night air, forming perfect conditions to reflect light down from over the horizon. When the weather's within a certain range, the lights appear; and at both locations, researchers have reliably and repeatably correlated appearance of the lights with control lights placed below the horizon. There are many such lights around the world.

At first glance, the Brown Mountain Lights have a lot of the same characteristics, and our first reaction might well be to classify it along with the others. But upon closer inspection, we have to say "Not so fast." For it seems that there are two distinct and very different manifestations of the Brown Mountain Lights.

To understand this, let's take a look at Brown Mountain itself. The main thing you'll notice is that it's hardly a mountain at all. It's in a region of foothills of the Appalachians, cleft by canyons and streams. The highest elevations in the area are hardly more than 850 meters, and one hill is hardly distinct from another. Brown Mountain is one such ridge. If it wasn't already named, it would scarcely occur to you to name it.

Moreover, Brown Mountain itself seems to have little direct relevance to the Lights. There are three most-often cited vantage points for viewing the Lights. The most popular is called Wiseman's View, about 13 kilometers west of Brown Mountain. Another is the Lost Cove Overlook, about 16 kilometers northwest of the mountain. Between the two is a Forest Service overlook about 8 kilometers away. From each of these vantage points, you look toward Brown Mountain, and it's hardly more than a smoky-colored smudge on the horizon. If you're lucky, somewhere in that intervening distance, you'll see the Lights. Brown Mountain itself is not the place to go to see them, nor does it appear to be the place where the Lights appear.

A National Park Service interpretive sign describing the Brown Mountain Lights

This brings us to what the lights look like and how they appear. To learn this, we'll go to Wiseman's View. Highway 105 runs along a ridge, quite a bit higher in elevation than Brown Mountain, and there's a scenic viewpoint turnoff with a parking lot. The lot overlooks the rocky Linville Gorge. Continue on foot down a paved path to a precarious overlook hanging over the gorge. Looking east across the gorge there's a dip in the opposite side, and through that dip you can see the far away ridge of Brown Mountain, and beyond that, the distant city lights of Winston-Salem. Historically, the lights were described as is printed on this United States Forest Service sign:

> The long, even-crested mtn. in the distance is Brown Mtn. From early times people have observed weird, wavering lights rise above this mtn., then dwindle and fade away.

If you tried to observe a hovering light above Brown Mountain from Wiseman's View, it would be lost in the city lights. Modern reports of the lights from Wiseman's View are, as mentioned earlier, a very different manifestation. People point their video cameras instead directly at the face of the hills across the gorge, looking not into the sky above the ridge, but straight at the hills. They report flickering lights under the trees, like people waving flashlights around. What's over there?

The right side of that dip looking across Linville Gorge is Table Rock, the most popular rock climbing destination in North Carolina. It's a dramatic rocky outcrop atop the hill across from Wiseman's View. Table Rock has its own parking lot only a couple hundred meters away, so it's not necessary for climbers to camp out on the slopes with their lanterns and flashlights. Nevertheless, it is approximately along the prominent hiking trails around Table Rock where the appearance of lights seem to be reported from Wiseman's View.

One explanation that's been offered, but not well received, is that these lights appearing on the face of the hills are reflections off of illegal moonshine stills, of the moon or other lights, or even the fires from the stills themselves. This is something that could be easily verified, but never has been. In any case, around the slopes of the state's most popular rock climbing destination, in plain view of the most famous overlook in the state, would not be a very clever hiding place for the shrewd brewer.

This type of light — a flicker directly visible on the hillside, as opposed to a probable refraction apparently hovering above the ridge — also contradicts local legend about the Lights' origin. According to modern locals, the Cherokee natives of the region believed 800 years ago that bereaved wives wandered the skies above the hills with lanterns, looking for the souls of their brave warrior husbands killed in battle. I say "supposedly" because I was unable to find any reference to such a belief outside of publications about the Brown Mountain Lights. That doesn't mean the Cherokee did not actually have such a legend; it only means I couldn't find it. I don't, however, recall the ancient Cherokee having lantern technology. Perhaps they meant torches, I don't know. But the retellings of this legend that I did find are unanimous in that the Brown Mountain Lights are specifically the type that appear in the sky above ridges, not flickering through the trees on the face of the hill.

This is also supported by one of the early "scientific" explanations of the Brown Mountain Lights. In 1771, the most prolific cartographer in Colonial America was the Dutchman John William Gerard de Brahm. He was appointed Surveyor General by the British and traveled throughout the Colonies constructing fortifications and bridges. He was also something of an amateur mystic. While in the southern United States, de Brahm is said to have published the following hypothesis (more on this in a moment):

> *"The mountains emit nitrous vapors which are borne by the wind and when laden winds meet each other the niter inflames, sulphurates and deteriorates."*

This is essentially the same "swamp gas" explanation used today to answer everything from ghost lights to UFOs. As a serious theory, it falls short of credible. Swamp gas never been found to spontaneously ignite in nature, as it would require a highly improbably mixture of gases in critical proportions. When these conditions have been created artificially in the lab, the gas burns bright bluish-green with a sudden pop, producing black smoke. Under no conditions does it burn slowly, or hover, or in any way resemble the reports of the Brown Mountain

Lights. Although it sounds scientific and convincing, the swamp gas hypothesis is almost certainly not the explanation.

By studying the oldest literature, we find that the Lights have been at least partially explained. Often cited as one of the earliest print references to the Brown Mountain Lights is a 1913 article from the *Charlotte Observer* entitled "No Explanation" that described how the light appears regularly at 9:30 or 10pm nearly every night. In 1922, the US Geological Survey produced a special report based on an exhaustive investigation of the Lights, and found that since 1909 a regularly scheduled locomotive on the plains beyond had been casting its headlight in precisely that direction every night at that same time. In fact it was due at a stop along there at 9:53pm every night. The locomotive's headlight was visible in a direct line of sight from a hill six miles beyond Brown Mountain, and certainly would have been from other locations as well. If it was not the source of the light described in the 1913 article, it would have been in direct competition with it; but witnesses reported one light, not two. Moreover, back in 1909 the USGS had actually made an earlier report in response to sightings of the Brown Mountain Light, and the investigator found that the sightings were all of this same train.

One member of the 1922 USGS team, a Mr. H. C. Martin, initially found geographic conditions in the region to be completely unsuitable to produce superior mirages of the sort that could refract distant lights from over the horizon and make them appear to hover in the air. But upon further investigation, he found that such refraction was taking place, nearly every night, above the basin now occupied by Lake James, and around which were numerous settlements with plenty of electric nights. From all observing stations, these lights were routinely seen above the horizon. The report noted:

> *"As the basin and its atmospheric conditions antedate the earliest settlement of the region, it is possible that even among the first settlers some favorably situated light may have at-*

tracted attention by seeming to flare and then diminish or go out."

This fits well with modern reports from Wiseman's View and the other overlooks that if you move up or down the mountainside, the lights vanish. Martin noted a narrow 3-4° angle in which such refracted lights would be visible.

The USGS also included historical research that found that it was not until 1910, when a Reverend C. E. Gregory moved into the area and began making reports, that the Lights became generally known. Even the belief that the Cherokee and Catawba natives had legends did not appear until about this time. Sightings that predate this period seem to all be apocryphal, with no actual print references known to exist. Even the often-cited 1771 report from de Brahm is suspect. It's always given out of context, and is, in fact, *misquoted*. De Brahm was not talking about any lights at all, he was giving his mystical and somewhat alchemy-centric opinions on how thunderstorms work and why the air is so clear in the Great Smoky Mountains. Here's what he actually said, and it was in his undated *Report of the General Survey in the Southern District of North America:*

> "Although these Mountains transpire through their Tops sulphurueaous and arsenical Sublimations, yet they are too light, as to precipitate so near their Sublimitories, but are carried away by the Winds to distant Regions. In a heavy Atmosphere, the nitrous Vapours are swallowed up through the Spiraculs of the Mountains, and thus the Country is cleared from their Corrosion; when the Atmosphere is light, these nitrous Vapours rise up to the arsenical and sulphureous (subliming through the Expiraculs of the Mountains), and when they meet with each other in Contact, the Niter inflames, vulgurates and detonates, whence the frequent Thunders, in which a most votalized Spirit of Niter ascends to purify and inspire the upper Air, and a phlogiston Regeneratum (the metallic Seed) descends to impregnate the Bowels of the Earth; and as all these Mountains form so many warm Athanors which draw and absorb, especially in foggy Seasons, all corrosive Effluvia along with the heavy Air through the

> *Registers (Spiracles) and thus cease not from that Perpetual Circulation of the Air, corroding Vapours are no sooner raised, than that they are immediately disposed of, consequently the Air in the Appalachian Mountains in extreamely pure and healthy."*

Taken in context, it's clear that de Brahm's quote has nothing whatsoever to do with the Brown Mountain Lights. This leaves us with no documentary evidence that the Lights existed at all prior to the arrival of electric lights and people in the area in the early 1900's.

So let's wrap up what we've learned about the two different manifestations of the Brown Mountain Lights. Regarding those that appear in the sky above a ridge, it's apparent that the 1922 USGS report solved it as described in the following conclusion. Today, nearly 90 years later, the lights are coming from different sources but this analysis probably still holds up:

> *"In summary it may be said that the Brown Mountain lights are clearly not of unusual nature or origin. About 47 percent of the lights that the writer was able to study instrumentally were due to automobile headlights, 33 percent to locomotive headlights, 10 percent to stationary lights, and 10 percent to brush fires."*

As for the lights appearing on the faces of the hills, we find there are no historical references to such a thing, and only a few recent YouTube videos and modern claims reporting it, in this age of LED flashlights, lanterns, headlamps, and iPhone screens. So I'm confident calling this one unexplained, but also not especially interesting or surprising.

It is all too often that we eagerly accept wild and sensational phenomena, which causes us to shut out the real science behind what's going on. I find real wonder in mirage refractions, and I find great excitement in such perfect solutions as the correlation of the locomotive with the 1909 Lights reports. This wonder and excitement are lost to those who replace science with sensationalism.

References & Further Reading

Caton, D. "See the Lights." *The Brown Mountain Lights.* Appalachian State University, 15 Mar. 2005. Web. 30 Sep. 2010. <http://www.brownmountainlights.org/>

De Brahm, J. *Report of the general survey in the Southern District of North America.* Columbia: University of South Carolina Press, 1971.

Editors. "No Explanation." *Charlotte Observer.* 24 Sep. 1913, Newspaper: 2.

Johnson, R. *Hiking North Carolina: A Guide to Nearly 500 of North Carolina's Greatest Hiking Trails.* New York: Globe Pequot, 2007. 149-150.

Mansfield, G. *Origin of the Brown Mountain Light in North Carolina.* Washington: United States Geological Survey, 1971.

Norman, M. *Haunted Homeland: A Definitive Collection of North American Ghost Stories.* New York: Macmillan, 2008. 308-310.

Sceurman, M., Moran, M. *Weird U.S.: Your Travel Guide to America's Local Legends and Best Kept Secrets.* New York: Sterling Publishing Company, Inc., 2009. 103.

Toomey, M. *Tennessee Encyclopedia of History and Culture.* Nashville: Tennessee Historical Society, 1998.

39. Boost Your Immune System (or Not)

Is "boosting your immune system" for real? Is that possible, and can you really buy it in a bottle?

In this chapter, we're going to point our skeptical eye at one of the most popular marketing gimmicks from the past few years: the sales of products and services with the claim that they will "boost your immune system". It sounds simple and desirable. Who wouldn't want a superpowered immune system capable of fighting off anything from a cold to cancer? Is such an ability really something you can buy in a bottle?

It's an easy claim to sell to people, because it's so clear and seems to make such obvious logical sense. The stronger your immune system, the greater its ability to fight disease. It sounds like it should be just like building muscle: A stronger bodybuilder can lift heavier weights, and a boosted immune system can fight off stronger diseases. Doesn't that sound right?

It may, but it's a completely invalid analogy. A healthy immune system is more accurately represented by a balanced teeter totter. If your immune system is compromised or otherwise weakened, one side of the teeter totter sags, and your body becomes more easily susceptible to infection. Conversely, if your immune system is overactive, the other side of the teeter totter sags, and the immune system attacks your own healthy tissues. This is what we call an autoimmune disease. Conditions like lupus, rheumatoid arthritis, psoriasis, and multiple sclerosis are all autoimmune diseases caused by "boosted", or overactive, immune systems. You're at your healthiest when the teeter totter of your immune system is balanced right in the center; neither too weak, nor too strong.

If you could boost your immune system, it would automatically and immediately be harmful.

So what do these companies mean when they claim their products boost the immune system? Fortunately for your body, they generally mean nothing at all. In recent years, the Federal Trade Commission and the Food and Drug Administration has been trying to crack down on products making unsubstantiated health claims, such as boosting the immune system. This is difficult for a number of reasons. First, "boosting your immune system" is medically meaningless; there is no such thing, so the assertion does not constitute a medical claim all by itself. You may recall that the supplement product Airborne was fined by the FTC and ordered to refund the money of everyone who had ever bought their product, but this was only because they went farther and specifically claimed their product could treat and prevent colds. Second, regulators are hopelessly outnumbered by the hordes of mail order and Internet businesses that can literally pop up overnight, to say nothing of the many well established companies like Airborne. Third, it's a very simple matter for such companies to subtly change their wording to make it even less specific, and thus escape prosecution. Today it's popular for products to say they "support a healthy immune system", and so they do, in the same way that any food or even breathing keeps your body alive and thus "supports" all its functions. They could just as honestly say their product supports body odor and aging.

Up until about ten years ago, nobody had invented the marketing term yet, so nobody ever thought to buy special supplements or specially grown produce to boost their immunity. Without such products, one wonders how the human race could have survived hundreds of thousands of years. Or even the 1980s or 1990s. Were we really less healthy then? Did we all truly have compromised immune systems?

You see, health is not the result of a superpowered immune system. Health is simply the absence of disease. Good health is the baseline. You can't be healthier than baseline. Once you're at the baseline, anything that happens to your immune system in either direction is bad. For a person in good health, who watches their diet and exercises, to walk into a smoothie store and order the special immunity-boosting supplement, would be harmful to their health. Would be, if that supplement actually did anything.

It would be easy for companies to demonstrate that their products work as advertised. The immune system is a surprisingly complex collection of structures and processes throughout the body. Many of these are types of cells that can be found in the blood. If a product actually boosted your immune system, it would have to increase the counts of one or more of these cell types. That's something that we could measure directly, and prove or disprove the claim. The problem with doing such a test is that it would be unethical, since you would have to give someone an imbalance likely to result in an autoimmune disease.

Let's take a quick look at what some of these systems are:

- The most obvious parts of your immune system are the external physical barriers: Your skin, saliva, tears, and processes such as coughing and sneezing. When we catch a cold, our immune system responds by increasing production of all of these responses. It's not the cold that gives you a running nose and makes you cough and sneeze, that's your immune system. Do you really want to boost that and walk around sneezing and drooling? Because that's what a "boosted" immune system means.

- Inflammation is another important immune response. Damaged cells release several types of triggering molecules that do such things as attract leukocytes or tell your body's blood vessels to dilate. These molecules can also hamper protein synthesis, which is intended to harm any viruses nearby that might have caused the damage. Inflammation is not a good thing. It's like a fireman spraying water on your house. You do it when you have to, you absolutely do not want to artificially stimulate inflammation when you don't have to.

- White blood cells, or leukocytes, are what most people think of when they hear "immune system". But what many people don't know is that there are many different types of leukocytes, not just those in our blood, but also other types in most of our other tissues as well. Leukocytes include macrophages, neutrophils, dendritic cells, mast cells, killer cells, and basophils, and others, all of which do different things.

All of those systems together comprise our "innate" immune systems, and they're just the half of it. We also have "adaptive" immune systems, and these are the systems that react to specific pathogens, multiply, and then become long-term guards against a recurrence of that same pathogen, becoming a sort of "memory" for your immune system. The adaptive immune system grows every time you challenge it with a specific germ, and it's also what reacts to a vaccine and becomes a prophylactic against a specific disease. The adaptive immune system is made up of special cells called lymphocytes, which include:

- Killer T cells, which do the dirty work, binding to and killing cells that match their specific receptor. Each killer T cell recognizes only one specific antigen, so we all have many, many, many different populations of killer T cells.

- Helper T cells are those which recognize pathogens, and express new T cell receptors for the killer T cells; in effect, creating new types of killer T cells designed to fight that specific pathogen.

- γδ (gamma delta) T cells perform a function similar to helper T cells, but are not necessarily triggered by new pathogens. Their function is pretty complex.

- B cells come between pathogens and helper T cells. They have a vast array of receptor proteins on their surfaces, and when a pathogen binds to one, the B cell divides millions of times. Each copy then finds a helper T cell to transfer the information about the new pathogen to killer T cells.

Those are the major components, but believe me, I've just given you an oversimplified 30,000-foot view of the immune system. Both halves, the innate and the adaptive, are comprised of many different components. Some act in concert, some act independently. There are many, many different ways in which parts of your immune system can be compromised, and addressing each of these deficiencies requires a different strategy. The notion that a single juice drink or supplement pill can "boost your immune system" is — to borrow a phrase — "so wrong it's not even wrong".

Since the function of the adaptive immune system is to react to challenges and develop new defenses, it can indeed be improved. Every time you catch a cold or get vaccinated, your adaptive immune system builds a new army of killer T cells ready to fight off a future recurrence of the same pathogen. There is no nutritional supplement, superfood, or mind/body/spirit technique that will do this for you. Those B cells only know which proteins to express by being attacked by specific disease agents.

The usual response that I hear to these arguments is that "immune boosting" products are simply trying to restore healthy immune function, since we're all walking around with compromised immune systems, because we eat badly and are obese and live in a toxic world. This is a familiar argument, and it's also easy to sell. It sounds like it makes sense. People do overeat, we love our prepared foods, and few of us take any special interest in the chemicals making up the objects in our daily lives. Has this truly resulted in compromised immune systems?

In fact, the opposite is true. Obese people generally have inflammation, which is an immune response. We catch colds and have no difficulty in producing symptoms. When we're exposed to irritating substances, we react with hives or itching or asthma, all of which are immune responses. Practically every one of us has some immune system response going on right now. The claim that living in our modern world has compromised our immune systems is measurably, and unambiguously, untrue.

There are real conditions in which immune systems can be compromised. These include primary immunodeficiencies, which are usually genetic and exist from birth, and require complex medical intervention; and acquired immunodeficiencies, usually resulting from disease, like AIDS, some cancers, even chemotherapy. A specific component of the immune system is affected and requires a specific treatment. Acquired immunodeficiency can result from malnutrition, but you have to be practically starving to death. It's the opposite problem from eating too many cheeseburgers.

Supplements, juices, or any products that claim to "boost your immune system" are frauds. They are for-profit solutions to a problem that does not exist and was invented by clever marketers to scare you into buying the products. Don't stand for anyone telling you that your balanced teeter totter can be brought into better balance by piling sandbags on one end.

References & Further Reading

Brain, M. "How Your Immune System Works." *Discovery Health.* Discovery Communications Inc., 1 Apr. 2000. Web. 8 Oct. 2010. <http://health.howstuffworks.com/human-body/systems/immune/immune-system.htm>

Crislip, M. "Boost Your Immune System?" *Science-Based Medicine.* Science-Based Medicine, 25 Sep. 2009. Web. 7 Oct. 2010. <http://www.sciencebasedmedicine.org/?p=1828>

Goldacre, B. "When it comes to a cold, you might as well try goat entrails." *The Guardian.* 22 Nov. 2008, Newspaper.

Hall, H. "Boost My Immune System? No Thanks!" *Skeptic.* 22 Mar. 2010, Volume 15, Number 4: 4-6.

Schindler, L. *Understanding the Immune System.* Washington, DC: U.S. Dept. of Health and Human Services, Public Health Service, National Institutes of Health, 1988.

Singh, S., Ernst, E. *Trick or Treatment.* New York: W. W. Norton, 2008.

Wallace, A. "An Epidemic of Fear: How Panicked Parents Skipping Shots Endangers Us All." *Wired.* 1 Nov. 2009, Volume 17, Number 11.

40. Speed Reading

Speed reading classes claim to be able to turbocharge your words per minute. Is this really possible?

We've all seen films of speed readers going through books nearly as fast as they can physically turn the pages. It's enough to make anyone envious. Who among us wouldn't love the ability to pick up any book, flip through its pages in just a few minutes, and then put it down in record time with nearly 100% retention? When I look at my vast stacks of unread books, the idea is certainly a compelling one. Fortunately for slow readers like myself, our demand-driven economy has responded with a product we can buy:  Classes and techniques purporting to be able to turbocharge our reading speeds to thousands of words per minute.

The most often cited speed reader is the late Kim Peek, the famous savant upon whom the *Rain Man* character was based. His mental abilities were so vast and varied that speed reading was hardly the most remarkable, yet it was still really something. He read two pages at a time, the left page with his left eye and the right page with his right eye. Estimates of his speed vary, but 10,000 words a minute is the number I found most often. Peek had a unique hardware arrangement driving this ability, though. He was born without a corpus callosum (the connection between the two brain hemispheres), and it's possible that his two hemispheres were able to process the pages he read in parallel. Kids, don't try this at home.

The most famous speed reader is probably John F. Kennedy, who spoke about it often and is said to have had his staff take Evelyn Wood speed reading classes. 1,200 words per minute is the number cited for Kennedy, however we'll look a little more closely at this in a few moments.

The *Guinness Book of World Records* does list a fastest reader, Howard Berg, who claimed 25,000 words a minute, nearly as fast as one can fan the pages of a book. Berg is best known for amazing stunts of speed reading and comprehension on television shows, including one with Kevin Trudeau who sold his speed reading course Mega Reading. But his claims were not without controversy. First, his TV stunts were incredible, but they never came near approaching 25,000 words a minute. Second, The Federal Trade Commission filed suit against him in 1990 for false and misleading advertising, after a blinded study found that none of his customers gained anywhere near as much as he said they would. Still, the fastest of those tested had quadrupled their speed to 800 words per minute.

How fast is 800 words per minute? It doesn't sound all that great compared to some of these other speeds. But apparently, 800 would be extremely fast for anyone without Kim Peek's hardware. Fast speeds require skimming, and comprehension drops off dramatically. It's always a trade-off. At 800, there's a massive loss of comprehension. To truly measure reading speed, we'd have to draw a line at some minimum acceptable level of comprehension.

Ronald Carver, author of the 1990 book *The Causes of High and Low Reading Achievement,* is one researcher who has done extensive testing of readers and reading speed, and thoroughly examined the various speed reading techniques and the actual improvement likely to be gained. One notable test he did pitted four groups of the fastest readers he could find against each other. The groups consisted of champion speed readers, fast college readers, successful professionals whose jobs required a lot of reading, and students who had scored highest on speed reading tests. Carver found that of his superstars, none could read faster than 600 words per minute with more than 75% retention of information.

Keith Rayner is a professor at the University of Massachusetts, Amherst and has studied this for a long time too. In fact, one of his papers is titled *Eye movements in reading and information processing: 20 years of research,* and he published that in 1993. Rayner has found that 95% of college level readers test between 200 and 400 words per minute, with the average right around 300. Very few people can read faster than 400 words per minute, and any gain would likely come with an unacceptable loss of comprehension.

So before you embark on a speed reading course, understand that knowledgeable professionals have devoted their careers to studying this, and have conclusively found that any gains you're likely to achieve are probably nowhere near the numbers printed in your class's marketing brochure, at least not without massive loss of retention. But let's take a look at the strategies that speed reading courses teach.

One of the basic goals is the elimination of subvocalization, claimed to be the thing that slows readers down the most. Subvocalization is the imagined pronunciation of every word we read. I do this a lot, and it limits my reading speed to virtually the same as my talking speed. Subvocalization is even accompanied by minute movements of the tongue and throat muscles. Nearly every speed reading class promises the elimination of subvocalization.

Here's the problem with that. You can't read without subvocalization. Carver and Rayner have both found that even the fastest readers all subvocalize. Even skimmers subvocalize key words. This is detectable, even among speed readers who think they don't do it, by the placement of electromagnetic sensors on the throat which pick up the faint nerve impulses sent to the muscles. Our brains just don't seem to be able to completely divorce reading from speaking. NASA has even built systems to pick up these impulses, using them to browse the web or potentially even control a spacecraft. Chuck Jorgensen, who ran a team at NASA in 2004 developing this system, said:

> *"Biological signals arise when reading or speaking to oneself with or without actual lip or facial movement. A person using the subvocal system thinks of phrases and talks to himself so quietly, it cannot be heard, but the tongue and vocal chords do receive speech signals from the brain."*

In fact, scientists have a term for reading in this way. They call it rauding, a combination of the words read and audio. To truly comprehend what your brain is seeing, nearly all of us must raud the words, fastest speed readers included. Fast readers need not be fast speakers; they simply have what's called a larger "recognition vocabulary". Rauding an unfamiliar word is subvocalized more slowly than a word already stored in our recognition vocabulary. We've learned that your recognition vocabulary, and thus your reading speed, can actually be improved; but the real technique is the opposite of what's taught in speed reading courses. Focus instead on reading comprehension. This will improve your recognition vocabulary, and you will probably begin to read faster.

Thus, elimination of subvocalization is a gimmicky claim. It sounds logical, and it's an easy sell. By skimming a text, you can subvocalize less of it, and you will comprehend less of it. Rauding the complete text is the only way to actually read it.

Another strategy taught in speed reading is special eye movements. These are usually things like reading lines backwards and forwards, and taking in several lines of text at a time. Again, this gimmick sounds like an attractive superpower to have, but it's counterintuitive to the way our brains actually process text. Those of us who aren't Kim Peek need serial input. Here's what's happening when you read. First, your eye lands on a point in a printed sentence. This is called a fixation, and it lasts (on average) a quarter of a second. Your eye then moves to the next fixation, and this movement is called a saccade, and takes a tenth of a second. After several saccades, your brain needs time to catch up and comprehend. This takes anywhere from a quarter to half a second. Half a second is a long time, and that's the rauding catching up with the saccades.

Is it possible to fixate once in a group of ten lines of text, and actually take it all in? Maybe, but only with a sufficient pause to comprehend before moving on. Speed reading teaches you to skip this pause, and thus your brain will not process the majority of what your eyes pass.

If we look back at the test that found Howard Berg's students improved to as much as 800 words a minute, we have to keep in mind that speed and comprehension are a trade-off. Whether 800 words a minute constitutes a passing score depends on what kind of comprehension threshold is set, and also what kind of text it was. When *The Straight Dope* administered its own speed reading tests, they found that people who had not read the texts at all often scored nearly as well on comprehension questions as the speed readers — when the text was general enough. In other words, it's very easy for professionals like Evelyn Wood or Howard Berg to control the conditions of the test to produce amazing results, good enough to impress television hosts, and to sell classes to laypeople.

So what about John F. Kennedy and his 1,200 words per minute? Kennedy biographer Richard Reeves looked into this. The 1,200 number comes from an off-the-cuff guess made to *Time* magazine's White House reporter. The reporter called the Evelyn Wood school where Kennedy had taken his speed reading class, but found that he had no score, as he'd never completed the class and actually been timed. But in what the reporter figured was a bit a PR posturing, the school told him that Kennedy "probably" read 700-800 words per minute. Carver's educated guess is that Kennedy likely read 500-600 words per minute, but may have been able to skim as fast as 1,000. So take the Kennedy claims with a grain of salt.

Test yourself at your normal reading speed, and you'll probably be surprised to learn that what you thought was slow is actually right in that normal range of around 300 words a minute. If you're much faster than that, you're among the few people with a highly developed recognition vocabulary. To improve this, stay away from gimmicky techniques that ignore the way the brain processes printed text, and focus on your comprehension. To read faster, concentrate on reading slower, and read more often.

References & Further Reading

Carroll, R. "Speed Reading." *The Skeptic's Dictionary.* Robert T. Carroll, 11 May 2000. Web. 14 Oct. 2010. <http://www.skepdic.com/speedreading.html>

Carver, R. *The Causes of High and Low Reading Achievement.* Mahway: L. Erlbaum Associates, 2000.

Just, M., Carpenter, P. *The Psychology of Reading and Language Comprehension.* Boston: Allyn and Bacon, 1987.

Noah, T. "The 1,000-Word Dash." *Slate.* Newsweek Interactive Co. LLC, 18 Feb. 2000. Web. 15 Oct. 2010. <http://www.slate.com/id/74766>

Rayner, K. "Eye movements in reading and information processing: 20 years of research." *Psychological Bulletin.* 1 Nov. 1998, Volume 124, Number 3: 372-422.

Reeves, R. *President Kennedy: Profile of Power.* New York: Simon & Schuster, 1993.

41. DDT: Secret Life of a Pesticide

Is DDT a killer of birds, a savior against malaria, or a little of each?

Put on your respirator and hazmat suit, because in this chapter we're pointing the skeptical eye at the claims on both sides of the DDT question. DDT is an insecticide in use since the 1930's. At first, its basic use was to kill mosquitos that transmit malaria, lice that transmit typhus, and other insect disease vectors like tsetse flies, at which DDT is extremely effective. It was so successful in World War II that its discoverer was awarded the Nobel Prize in medicine in 1948. Subsequently it was used in agriculture to protect crops from a variety of pests, and once again, it's highly effective in doing so. But a few decades later, DDT became a two-sided issue, with detractors pointing to health effects on humans and animals; most notably, eggshell thinning in various bird species, and a number of potentially severe health effects in humans. In response to these concerns, DDT has now been banned for the most part in many countries. But the controversy continues. While the ban has been credited with the rebound of bird species, it has also been criticized as overzealous, with many now saying the detrimental effects were overblown and did not outweigh the many lives saved from malaria in the third world. It is in fact making a comeback, with production increasing today in India, China, and North Korea, for both agricultural and anti-malaria uses.

And so we ask the question: Is one side completely wrong and one side completely right, or do we equivocate and conclude that DDT has its place, albeit a limited one?

DDT is dichlorodiphenyltrichloroethane. It's a completely synthetic compound that does not exist in nature. It's a white, powdery, waxy substance that's hydrophobic: It doesn't dissolve in water and so does not contaminate it, but readily dissolves in solvents and oils. It's applied as a white smoky mist. DDT kills insects by chemically enhancing the electrical connections between their neurons, short-circuiting them into spasms and death. DDT's hydrophobic nature is both a blessing and a curse. It can't contaminate water sources, which is good; but it also doesn't get dissolved away by them and diluted into virtual nothingness, so it hangs around for a long time.

DDT probably never would have been banned if it were not for the 1962 publication of *Silent Spring*, the title of which alluded to dead birds. Author Rachel Carson was a much beloved nature writer who died only two years after the book came out. She'd been a marine biologist for the U.S. Fish & Wildlife Service (then called the Bureau of Fisheries), but was able to retire with the publication of a trilogy of books about the sea. All became bestsellers in the 1950s, with the public enamored by her poetic presentation of all things pertaining to beaches, islands, the deep sea, and the creatures living there. Following this trilogy, her writing turned toward environmental issues and became increasingly critical of industry, government, and the effect of humans on the planet. *Silent Spring* was serialized in *The New Yorker* before its publication, and it was probably the most scathing of her works, though beautifully written. It charged DDT with being a health hazard and with widespread environmental destruction, particularly to bird populations, and was

A worker spraying DDT

unquestionably the turning point which resulted in DDT's bans in the United States and other countries. In fact, as one Environmental Protection Agency writer put it:

> "*Silent Spring* played in the history of environmentalism roughly the same role that Uncle Tom's Cabin *played in the abolitionist movement.*"

Rachel Carson's list of posthumous honors is a long one, showing what deep roots *Silent Spring* thrust not only into the environmental movement, but also into the public psyche. President Jimmy Carter awarded her the Presidential Medal of Freedom. She appeared on postage stamps in several countries. A bridge in Pittsburgh is named after her, as is a government building in Harrisburg. The number of schools, parks, nature refuges, scholarships and scholarly prizes named for Rachel Carson would fill a page.

Silent Spring's principal popular effect was to brand DDT as dangerous to bird populations. The mechanism by which DDT does this is now largely, but not completely, understood. In summary, it interferes with the delivery of calcium carbonate to the eggshell gland, and the eggs that are laid have thinner shells. Shells that are too thin can lead to the death of the embryo. This eggshell thinning is the primary environmental concern over DDT.

It's been about five decades since *Silent Spring* was published, and we've learned a lot in those years. One thing we've learned is that DDT is only one of many causes of eggshell thinning. Other culprits include lead and mercury toxicity, oil, phosphorus and calcium deficiency, and dehydration. Perhaps most significantly, birds in captivity in order to undergo testing are under stress, and this stress alone is enough to produce eggshell thinning. Although DDT's mechanism for eggshell thinning is plausible, many studies throughout the 1970's and 1980's failed to correlate such thinning with high levels of DDT, even extremely high levels. Other studies have confirmed Rachel Carson's findings. My own conclusion based on a review is that there probably is a correlation, but it's not a strong one; and at best it's only one of many causes. Whether DDT is used or not would probably not have a large impact on bird populations.

But despite the likelihood that it would have some impact, it's now known that the species Rachel Carson focused on (most notably bald eagles) were already in massive decline from unrelated pressures even before DDT's introduction. Habitat loss and hunting had been, by far, the greater causes of bald eagle deaths. Hunting had reduced the populations to just a few hundred nesting pairs in the mountains, and lowland eagles were already gone from habitat loss. Rachel Carson did not ignore these issues in her book, but the popular perception that banning DDT was all that was needed to magically restore bald eagle populations was naïve. In the end, it was the Bald Eagle Protection Act and the bird's 1967 placement on the endangered species list, combined with increased penalties for poaching, that ultimately led to the bald eagle's successful return to remaining habitats.

Brown pelicans are another species often cited as having been decimated by DDT use in the United States, along the Gulf coast and in California. Massive declines were indeed correlated with DDT use, but it may have been a coincidence in each case. Along the Gulf coast, hunting by angry fishermen had reduced the pelican population in Texas from 5,000 annual births to just 200 in 1941. DDT certainly didn't help; but it was another case where the bird populations would have dropped sharply whether DDT was in the picture or not.

Of course, it would be completely wrong to overlook DDT's potential for causing harm simply because there are other things that cause harm too. All we can do is our best to quantify exactly what the risk really is, and then the decision to ban or not to ban becomes a cost/benefit analysis, which is no longer a science question. Everyone has the right to their own opinion on what's most important, and in the United States, we chose the birds.

Silent Spring's legacy may have been good for the birds, but not so much for human populations in the third world. DDT is one of the most effective pesticides ever discovered for fighting malaria. Although DDT remains legal for insecticide use in most areas where malaria is a major killer, the money for fighting the mosquitos often comes from donors in wealthy countries like the United States. Such wealthy donors have often had little personal exposure to the issues, and can sometimes have a skewed perspective when it comes to bald eagle eggshells in the United States versus the deaths of children in Mozambique. Writing in the *Nature Medicine* journal, malaria advocate Prof. Amir Attaran criticized American environmental groups for opposing the public health exceptions of DDT bans:

> *"Environmentalists in rich, developed countries gain nothing from DDT, and thus small risks felt at home loom larger than health benefits for the poor tropics. More than 200 environmental groups, including Greenpeace, Physicians for Social Responsibility and the World Wildlife Fund, actively condemn DDT."*

As a result of these pressures, many donations now coming from wealthy nations are now contingent upon DDT not being used, which leaves the poor nations with fewer options, often too expensive and less effective, and children die. Up to three million people die of malaria each year, most of them in Africa. DDT, while it does have environmental and health concerns like all pesticides, is not known to have ever killed anyone. If we shelve our most effective tools hoping that something perfect will come along that has no potential downside, we'll wait forever, and thousands will continue dying every day. These are the cases where wealthy environmental groups appear to do their best to justify their elitist stereotype, at the expense of brown people. (The World Health Organization's ban on DDT does include limited exemptions for malaria control in many regions, but money for its use still often depends on qualified foreign aid. In Africa, the exemption allows indoor use only, like wearing armor on half your body.)

Rachel Carson absolutely acknowledged DDT's importance to fighting malaria, but was quick to point out another downside: acquired resistance. After six or seven years, mosquito populations develop resistance to DDT. However, this is the case with all pesticides, it is not a reason to avoid DDT per se. Moreover, we've since learned that it is still effective against resistant mosquitos, only a little less so. Susceptibility in resistant strains goes down to 63%, as opposed to 87% in non-resistant strains.

Even among resistant mosquitos, DDT is an exceptionally effective repellent. Houses treated with DDT are avoided by all mosquitos, resistant or not.

But like all synthetic chemicals, DDT has been blamed for virtually any human illness imaginable. Some say it causes cancer, diabetes, hyperthyroidism, loss of fertility, that it functions as an endocrine disruptor, and more. The World Health Organization classifies it only as "moderately hazardous", and in response to all the wildly conflicting studies of its cancer-causing effects in animal tests, the U.S. Environmental Protection Agency classifies it as a "probable carcinogen". The claims that DDT definitely causes cancer or anything else are not supported by the data, but obviously it's a risky compound that we don't want to expose anyone to if we don't have to. And so, again, we're outside of science questions, and down to risk assessment.

DDT does have its place, and its current usage is probably not too far off of what it should be. The exception is Africa where DDT's upside far outweighs the down, and my opinion is that donors should relax their restrictions against it, and leave those decisions to the experts on the front lines in Africa. For much of the rest of the world, DDT has largely been supplanted by newer and better agricultural pesticides, and there's insufficient reason to put collateral species under pressure. A scientific review nearly always produces better focused policy, and our DDT policy is definitely due for a tune-up.

References & Further Reading

Attaran, A., Roberts, D., Curtis, C., Kilama, W. "Balancing risks on the backs of the poor." *Nature Medicine.* 1 Jan. 2000, Number 6: 729-731.

Campbell, L. *Endangered and threatened animals of Texas: Their life history and management.* Austin: Texas Parks & Wildlife, Resource Protection Division, Endangered Resources Branch, 1995. 58.

Carson, R. *Silent Spring.* New York: Fawcett Crest, 1964.

Edwards, J., Milloy, S. "100 things you should know about DDT." *JunkScience.com.* Steven J. Milloy, 7 Jan. 2007. Web. 2 Nov. 2010. <http://www.junkscience.com/ddtfaq.html>

EPA. "DDT - A Brief History and Status." *Pesticides: Topical & Chemical Fact Sheets.* US Environmental Protection Agency, 16 Jan. 2008. Web. 9 Dec. 2010. <http://www.epa.gov/pesticides/factsheets/chemicals/ddt-brief-history-status.htm>

Gates Foundation. "Our Work in Neglected Diseases: Visceral Leishmaniasis, Guinea Worm, Rabies - Overview & Approach." *Bill & Melinda Gates Foundation.* Bill & Melinda Gates Foundation, 14 Jun. 2010. Web. 30 Oct. 2010. <http://www.gatesfoundation.org/topics/Pages/neglected-diseases.aspx>

Gladwell, M. "The Mosquito Killer." *The New Yorker.* 2 Jul. 2001, Annals of Public Health: 42.

Miller, H. "Utterly Repellent." *Forbes.com.* Forbes.com LLC, 24 Feb. 2010. Web. 29 Oct. 2010. <http://www.forbes.com/2010/02/24/bill-gates-malaria-vaccines-opinions-contributors-henry-i-miller.html>

Stokstad, E. "Can the Bald Eagle Still Soar After It Is Delisted?" *Science.* 22 Jun. 2007, Volume 316, Number 5832: 1689-1690.

42. The Mystery of STENDEC

What was the significance of this mysterious final transmission of an airliner just before its crash?

It was a story borne out all too often in the annals of aviation disasters. An aircraft finds itself off-course and in the clouds with zero visibility, and worse, surrounded by mountain peaks that can't be seen. That's what happened on August 2, 1947, in the Andes Mountains of western Argentina. An Avro Lancastrian passenger plane of British South American Airways with 11 people on board struck a mountainside in zero visibility while descending toward what it thought was Chile. All aboard were presumed killed, with the crash not even being confirmed until more than fifty years later when its wreckage was finally found.

By themselves, the basic facts of what happened to British South American Flight CS-59 are tragic but not especially mysterious. The cause of the crash was a controlled flight into terrain. Nothing was broken, nothing was wrong, and it's unlikely the crew even saw the crash coming until the near-vertical cliff near the summit of Tupungato Mountain appeared through the mist. A second later, the Lancastrian was in a billion torn pieces, sliding down the face in an avalanche of its own making. Those pieces began to emerge from the bottom of the glacier, bit by bit, around the year 2000.

The mysterious part is what came over the radio just before the crash. In those days, long distance communication was by Morse code. The crew had been making regular hourly reports of its speed, position, and altitude, and all was well. At 5:41pm, four minutes before their anticipated landing time, the radio operator, Dennis Harmer, sent their ETA as 5:45pm and concluded it with the phrase STENDEC:

... ¯ . ¯. ¯... ¯.¯.

The Chilean Air Force ground controller asked him to repeat that, and Harmer sent it twice more, STENDEC, STENDEC.

And that was the last anyone ever heard of Flight CS-59. The mystery of the word STENDEC took its place among the great unsolved cases so beloved in the lore of urban legendry.

The Avro Lancastrian began its life as a British Lancaster bomber in World War II. Lancasters had four Rolls Royce Merlin engines, the front-line combat engine that powered the latest Spitfire and Mustang fighters. With some 5,000 horsepower on tap, Lancastrians were in demand for civilian applications after the war, such as the

Avro Lancastrian

route over the Andes. The crew and passengers of Flight CS-59 had in fact just crossed the Atlantic from London on board an Avro York, a slightly different variant of the Lancaster; then changed planes to the Lancastrian for the final leg.

The JIAAC (the Argentinian Civil Aviation Accident Investigation Board) found the probable cause of the crash, but only after examining the wreckage upon its 2000 discovery. In CS-59's day, navigation was done by dead reckoning, making time/speed/distance calculations, as there were no radio beacons to help determine position. Flying east to west across South America, a plane must fight the jet stream head on. If the jet stream is faster than expected, dead reckoning calculations will lead you to believe that you've gone further than you actually have. Flight CS-59 had been in the air long enough to clear the Andes, and with no visual cues, began its descent into Chile, down into the clouds. Unfortunately, they descended right into a collision course with Tupungato. Once they entered its wind shadow, they fell victim to a phenomenon called a lee wave, part of which involves a downdraft over the face of the mountain, and at that point a crash became unavoidable. Without visibility, they almost certainly never saw it coming, and couldn't have done anything about it even if they had. These are precisely the same conditions that caused the 1972 crash of the Uruguayan Rugby team that inspired the book and movie *Alive*.

CS-59's crew were all experienced Royal Air Force pilots who had flown against the Germans in World War II. The pilot, Reginald Cook, had flown more than 90 combat missions. Cook's first officer Norman Hilton-Cook and second officer Donald Checklin each had over 2,000 flight hours. Dennis Harmer had served as a radio operator for three years during the war, and for over 600 hours for the airline. Among the crew, they had some 30 crossings of the Andes Mountains, though this was Cook's first time as captain.

Certainly such a crew would have been aware of the jet stream and lee waves, but none had ever negotiated the Andes before without an experienced captain. Although the cause of the crash was navigational error coupled with pilot error, the crew did about as well as anyone could have done, given the technology of the time, and the weather conditions nature thrust them into.

But the 2000 finding of the true cause of the crash has not stemmed the speculation about what significance STENDEC had. Theories even extended to UFOs, with some people proposing the cryptic message must have been some kind of warning about the plane's abduction by a giant alien craft. Even aviation medical experts have gotten into the fray, suggesting that the crew may have been suffering from hypoxia. Lancastrians were not pressurized but the crew did wear oxygen masks, and perhaps Harmer was not wearing his or it had malfunctioned. But if this was the reason for his garbled transmission, it seems unlikely that he could have repeated it verbatim, twice, when asked by the Chilean controller.

All kinds of people have made all kinds of guesses about what STENDEC might have meant. None of them are very convincing. Most interesting is that the letters are an anagram of DESCENT, and Flight CS-59 was certainly on its descent. Did Harmer simply miskey the letters of DESCENT? It seems to be quite a stretch that such an experienced operator would have made the exact same extreme misspelling three times in a row. Moreover, it would have been a totally random place in the transmission to throw in this word. Nobody ever did that.

It's also been proposed that Harmer miskeyed the Aircraft's name, which was *Star Dust* (all of British South American Airways' aircraft were given two-word names beginning with *Star*). This possible misspelling has been a popular theory, so much so that most stories you'll find about Flight CS-59 are titled *The Mystery of the Star Dust* or some such thing. However, in reality, nobody at the time ever referred to the aircraft or the flight by the unofficial name painted on it by the airline. Probably the crew was scarcely even aware of the plane's name. In radio transmissions of the day, planes were only ever addressed by registration number, in this case G-AGWH. And, once again, it would have been completely random and uncalled for to tack the plane's name onto the end of a routine transmission.

Most other theories are that STENDEC was an acronym of the first letters of words in some sentence, and various nominations have been proposed. None of them make sense, because no radio operator would ever communicate in such a way, and certainly not repeat it twice more when told that it was not understood.

One fairly bad match is "Santiago tower, now descending entering cloud", which is true but no sane radio operator should expect such an acronym to be understood. Another is *"Star Dust* tank empty no diesel expecting crash", which has five problems: Harmer would not have identified the aircraft by its unofficial name; *Star Dust* was two words, not one; Merlin engines do not burn diesel; the plane was not low on fuel; and Harmer was experienced enough to know there would be no reasonable expectation of anyone comprehending such an acronym.

Morse code experts have searched for a solution within the subtleties of hand-keyed code. There is an attention-getting signal consisting of three dots followed by a dash:

...-

...and these are the same as STENDEC's first two letters, except that they have a space:

... -

Morse code also uses something called prosigns, which are two letters run together without a space. The prosign AR, which means "End of message", looks like this:

.-.-.

...and that's the same as STENDEC's final two letters, except that, again, they have a short space after the first dot:

. -.-.

So, theoretically, STENDEC could mean ATTENTION - END - END OF MESSAGE. Now, this sounds fairly reasonable. And it could be, except that it wouldn't make much sense for Harmer to send this. STENDEC followed his transmission of their estimated arrival time, so it would be rather strange for him to insert the ATTENTION signal in the middle of a message. Similarly, it would be redundant to say END and then transmit END OF MESSAGE. None of Harmer's other transmissions from the flight did this, and no radio operators of the day ever did, so far as I could find. I think we can call this the best effort from the Morse code experts, and it shows that there is probably not a good answer to STENDEC to be found within the imperfections of hand keying.

Is it really necessary to assign some profound significance to STENDEC? What about all the other meaningless, garbled, or unclear radio transmissions that happen every day, that nobody remembers because they're not followed by a fatal crash? I've flown a lot, and I think every time I go up I hear someone say something I can't make out. I've no reason to suspect it was any different in the days of Morse code. Who knows what Harmer meant to say? Most likely, he was about to continue with what would have been a meaningful transmission when it was cut short by the crash. Most of these guys were in their twenties and Harmer could have even just been horsing around. The bottom line is that the cause of the crash had nothing to do with anything Harmer sent or didn't send on the radio. It would have crashed anyway, and chances are that if Harmer had lived five minutes longer he and the Chilean air traffic controller would have connected a little better, and the miscommunication never would have been remembered.

I see no reason to link the miskeyed, misread, or made-up transmission with the crash. The only thing connecting them is that they happened within a few minutes of each other, but the crash is simply not missing any explanations requiring an interpretation of STENDEC. Consider it an unsolved little tidbit from history, but one whose significance is almost certainly limited to the interest it holds as a puzzle.

References & Further Reading

Dinkins, R. & J. "CW Operating Aids." *AC6V's Amateur Radio & DX Reference Guide.* Rod Dinkins & Jeff Dinkins, AC6V, 17 Jan. 2008. Web. 29 Oct. 2010. <http://www.ac6v.com/morseaids.htm>

Editors. "Lost plane found in Andes." *BBC News.* British Broadcasting Corporation, 26 Jan. 2000. Web. 29 Oct. 2010. <http://news.bbc.co.uk/2/hi/uk_news/618829.stm>

NOVA Online. "1947 Official Accident Report." *Vanished! Teacher's Guide and Resources.* WGBH Science Unit, 30 Jan. 2001. Web. 28 Oct. 2010. <http://www.pbs.org/wgbh/nova/vanished/sten_report.html>

Rayner, J. *Star Dust Falling: The Story of the Plane that Vanished.* New York: Doubleday, 2002.

Ruffin, S. *Aviation's Most Wanted: The Top 10 Book of Winged Wonders, Lucky Landings.* Herndon: Brassey's, 2005. 26.

Taylor, M. *Jane's Encyclopedia of Aviation, Volume 2.* New York: Crescent Books, 1980. 257.

43. THE SOUTH ATLANTIC ANOMALY

Was this mysterious region in the south Atlantic responsible for the crash of Air France flight 447?

You've heard of the Bermuda Triangle, you may have even heard of its Pacific counterpart the Devil's Sear near Japan. But did you know about the South Atlantic Anomaly? It's a region over South America and the south Atlantic Ocean centered at about 30 degrees West and 30 degrees South, and it's big, a little larger than the entire United States. Some say it's a danger to anything entering it, and they point most specifically to the crash of Air France flight 447 in June of 2009, in which 228 people died. Whereas the Bermuda Triangle and the Devil's Sea are merely lines drawn on a map, the South Atlantic Anomaly is a measurable physical presence. Today, Skeptoid takes a look at what it is, what kind of danger it presents, and whether it really brought down Flight 447.

To understand the South Atlantic Anomaly, let's start by taking a quick look at the Earth's magnetic fields and the Van Allen radiation belts.

We all know that the Earth's magnetic axis is not the same as its rotational axis. As the Earth's molten, ferromagnetic liquid core churns, it generates a magnetic field, and the north-south axis of this field is tilted about 16° from the rotational axis. The north magnetic pole is in the north Canadian islands, but it moves around a lot, and it's currently headed northwest at about 64 km per year.

Here's the part that many people don't know. While the north magnetic pole is about 7° from the north rotational pole, the south magnetic pole is about 25° from the south rotational pole. A line drawn from the north magnetic pole to the south does not pass through the center of the Earth. Our magnetic field is torus shaped, like a giant donut around the earth. But it's not only tilted, it's also pulled to one side, such that one inner surface of the donut is more squished up against the side of the earth than the other. It's this offset that causes the South Atlantic Anomaly to be at just one spot on the Earth.

The anomaly itself consists of radiation. It is, in fact, the inner Van Allen radiation belt. If you picture the Earth as a circle enclosed within double parentheses, then bent those parentheses sharply toward it, you can visualize the shape of the Van Allen belt. The inner belt is shaped about like the letter C, while the outer belt is longer and almost completely encloses the inner belt and bends in sharply toward the magnetic poles. Since this whole field is tilted and offset to one side from the Earth, there's one point where the inner Van Allen belt almost touches the surface. This is exactly what the South Atlantic Anomaly is.

The belts are shaped by the solar wind, and as the Earth spins within them, the offset causes the whole system to wobble and everything is always in motion. This constant stirring is one reason the Northern Lights aren't static; they dance around. Correspondingly, the South Atlantic Anomaly observes a daily cycle, growing and shrinking and changing intensity, strongest at solar noon each day and weakest around solar midnight. One result of the solar wind hitting the magnetic field is the creation of a cavity within the field called a Chapman-Ferraro Cavity, and charged particles are trapped and collect within this cavity. These trapped particles are what the inner and outer Van Allen belts physically consist of.

The inner Van Allen belt, where it comes closest to Earth at the South Atlantic Anomaly, is a cloud of energetic protons. Their energies range from about 100 keV (100,000 electron volts), which is enough to penetrate half a millimeter of lead, all the way up to about 400 MeV (400,000,000 electron volts), enough to penetrate 143 mm of lead. You would not want to stay there very long. In fact this is one of the arguments made by the conspiracy theorists who say we never went to the Moon. But the irregular shape of the Van Allen belts relative to the Earth provided part of the solution for the Apollo astronauts, in the form of a gap large enough to squeak through with only minimal exposure. The rest of the solution lay in Apollo's speed, and the astronauts cleared the belts in only one hour.

Many satellites are not so lucky though, and have to pass through the South Atlantic Anomaly frequently. Some 200 satellites currently have to deal with this. Many, like the Hubble Space Telescope, turn off certain sensitive sensors whenever they pass through. Electronics on such craft that are vulnerable must be shielded. The International Space Station also has this problem, but since it completes an entire orbit every 90 minutes, its exposure time is very brief. Also only a few of its orbits pass through the Anomaly. The total radiation exposure received by space station residents is within safe levels, though they've reported odd things like computers crashing when they pass through the Anomaly. Scary stuff.

We know that the South Atlantic Anomaly is hazardous to electronic equipment and to humans who spend time inside it. We know that it dips down close to the Earth. So if we add these together, it suddenly sounds plausible that Flight 447 might have been taken down by it. Right?

Well, all by itself, wrong. Although the Anomaly is a dangerous place, its edges are pretty well defined. The closest it ever gets to the Earth's surface is about 200 km, and at that height it's very small. The highest commercial aircraft might ever get is about 15 km, and usually lower. Flight 447 was absolutely not, no way, impossible, *not* in the South Atlantic Anomaly, either when it crashed, or at any other time. The Anomaly is in space, it is not in the atmosphere; and you're completely safe if you're below it and not in it. Your computers will not crash, your electronics will not fry, and you will not be baked by radiation. You'll never even know it's up there.

Consider also that half of South America, including nearly all of Brazil, is directly underneath the South Atlantic Anomaly. Aircraft fly safely throughout South America all day every day. Computers, hospitals, and data centers operate here trouble free year round. There is simply no evidence, nor any plausible hypothetical argument, to suspect that the relative nearness of the Van Allen belts posed any danger at all to Flight 447.

Moreover, Flight 447 had already cleared the area below the Anomaly when it disappeared. Its last reported position was some 1,000 kilometers beyond the Anomaly's northern reaches. It made it past the Anomaly, and then continued flying safely northeast for nearly two hours before whatever happened happened.

Parts of Air France 447 being recovered

So how do we explain that little detail, the crashed airplane?

The crash was investigated by the BEA, France's version of the American NTSB, the Bureau of Investigations and Analysis for the Safety of Civil Aviation. The aircraft's flight data recorders were recovered in 2011. At about the same time that it disappeared from radar, the plane's systems send a burst of automated error messages to Airbus for maintenance purposes. These messages were almost all about inconsistencies between the three pitot probes that measure airspeed, almost certainly indicating that they were icing up. The plane was flying through storm clouds, though not a strong storm and not one to normally be concerned about, but evidently it was enough to ice up the airspeed sensors. In fact, a number of such automated error messages from other planes had prompted Air France to update the pitot probes on its Airbus 330s and 340s, and the replacement probes for this particular aircraft had been received by Air France six days before the crash, and were due to be installed.

The aircraft then stalled at 38,000 feet, and plunged to the ocean in only three and a half minutes.

Analysis of the debris and the bodies showed that the plane was in cruise mode and that no preparations for a crash had been made. It bellyflopped flat on the water, hard enough to kill the victims and destroy the aircraft. The BEA has determined that insufficient information exists to determine the exact cause of the crash, and they are in fact making plans to again try to recover the flight data recorders.

A number of independent researchers have not given up on trying to pin the blame for the crash on solar radiation. On May 31, earlier in the day, Sunspot 1019 appeared. The principal danger of sunspots is a wave of ejected high-energy protons, which start arriving at Earth about 15 minutes after a solar flare event and continue for many hours. Could a burst of solar radiation from Sunspot 1019 have knocked down Flight 447?

It is, absolutely, a possibility. But it's not as strong a candidate as it may initially seem. According to NOAA's space weather report for the period, Sunspot 1019 produced no flares on May 31, and no proton events at all were observed. Flux was reported to be at normal levels, and geomagnetic activity was described as quiet.

This reduces the possibility that solar or cosmic radiation caused the crash to pure speculation. It's improbable that a rogue particle or two would have caused a system error as specific as multiple inconsistencies reported by the airspeed sensors, especially given that icing from the storm cloud was a far more probable culprit. How this translated into the subsequent systemwide failures and crash is still unknown, but there's no evidence at this point that any external causes were at fault.

Perhaps a future recovery of the flight data recorders will shed additional light, but for now, there don't seem to be any questions about Flight 447 that can only be answered by the South Atlantic Anomaly. It's there, and it's real, and it does pose problems for spacecraft and astronauts passing through it, but all the information we have now indicates there's no reason for people on the ground or in the air beneath it to be concerned.

REFERENCES & FURTHER READING

Bailey, J. *Biomedical Results of Apollo.* Washington, DC: Scientific and Technical Information Office, National Aeronautics and Space Administration, 1975. Chapter 3.

BEA. *Update on the Investigation into the Accident to flight AF 447 on 1st June 2009.* Paris: Bureau d'Enquêtes et d'Analyses, 2009.

Goldhammer, L. *Proton spectrum characteristics of the 7 July 1966 solarflare.* Pocatello: Idaho State University, 1967.

Hope, F. "Did a solar flare knock Air France 447 out of the sky?" *Future News Today.* Frank Hope, 21 Jun. 2009. Web. 15 Nov. 2010. <http://futurenewstoday.blogspot.com/2009/06/did-solar-flare-knock-air-france-447.html>

IPS Radio & Space Services. "The Most Powerful Solar Flares Ever Recorded." *SpaceWeather.com.* Dr. Tony Phillips, 20 Apr. 2001. Web. 14 Nov. 2010.
<http://www.spaceweather.com/solarflares/topflares.html>

NOAA. "Preliminary Report and Forecast of Solar Geophysical Data." *Space Weather Prediction Center.* 2 Jun. 2009, Number 1761: 1.

Walt, M. *Introduction to Geomagnetically Trapped Radiation.* New York: Cambridge University Press, 1994.

44. IQ Testing

How valid are IQ tests, what do they really measure, and where do you fit in?

Don't you just *love* the idea that your level of intelligence can be boiled down to a single number, and ranked along with those of all the other dummies in the world? You may have taken an IQ test in the past, and may even know your score. It's an unfortunate fact of statistics that half the people walking around are below average intelligence — the way the tests are scored assures that 100 is both the median and the average — and sometimes we question the value of force-ranking ourselves, and assigning so much cultural significance and stigma to it, based on one narrow metric. IQ tests look like the ideal place to point Skeptoid's skeptical eye.

There are a number of obvious apparent criticisms of the idea of ranking everyone with a single number that purports to encompass how intelligent they are. Some people are "book smart" but with no "common sense", and some are the opposite. Some people have high or low creativity or humor, but may ace all their tests in school or fail them. Each of us is complex, with many strengths and weaknesses, aptitudes and preferences, and it seems that any one number purporting to quantify our intelligence must be grossly misleading in every case.

There are even obvious criticisms of the tests themselves. There are a number of different IQ tests in use, and it's well established that the same people will score differently on the various tests: I might get a higher score than you on one test, while you outscore me on another. Critics often point out that any IQ test is necessarily skewed toward a particular cultural frame of reference, making it unfair to measure someone from Africa using a test developed in Denmark (for example).

These basic criticisms are answered by a closer study of what IQ tests actually purport to measure. They've got nothing to do with "book smarts" and are intended to have no cultural relevance. The tests measure only your intelligence. There are as many different definitions of intelligence as there are psychologists, but we can extract some common themes from the definitions offered by those who have played the biggest roles in developing these tests. Generally speaking, your intelligence is your problem solving and reasoning ability. It encompasses learning, planning, and understanding.

IQ testing has an ominous history. It originally grew out of the eugenics movement in the United States around the turn of the twentieth century. The basic idea of eugenics was to identify desirable traits, such as intelligence, health, and even financial success, and to increase birth rates among such people. At the same time, birth rates among people with negative traits such as lower intelligence, criminal behavior, poverty, and illness, would be discouraged. When it was discovered that heredity played a large role in some mental illnesses, forced sterilization was imposed upon mental patients in some states in an effort to breed such traits out of the population. According to most counts, some 64,000 mentally ill Americans were sterilized until the practice was finally terminated in the 1960s. In the Nuremberg Trials, it was revealed that the Nazis considered the American program so effective that it was the inspiration for the Nazis' forced sterilization of some 450,000 people.

The father of eugenics was the Englishman Sir Francis Galton, a cousin of Charles Darwin. Over the course of Galton's varied and productive career, he not only codified the science of eugenics but also pioneered psychometry as a tool for measuring people's intelligence and determining whether it would be best for them to breed or not. Galton coined the phrase *nature versus nurture* and identified the trend of *regression towards the mean*, though his original term for this was *re-*

version towards mediocrity. So long as unintelligent people were allowed to reproduce freely, mankind could never rise above its native mediocrity.

A tool for quantitatively identifying mental retardation was needed by American eugenicists, and so they turned to two French researchers, Alfred Binet and Théodore Simon, who had developed the Binet-Simon test as a way of identifying French schoolchildren who needed special assistance. Binet-Simon did not ask questions about general knowledge, instead it imposed a diverse system of tasks, from simple physical tests to memory puzzles. The resulting score was expressed as the mental age.

Lewis Ternan, a psychologist from Stanford University, translated and improved the test in 1910, and it became known as the Stanford-Binet. The result was your Intelligence Quotient, a quotient of your mental age divided by your chronological age. If you were 10 years old but had the reasoning ability of a 15-year-old, your IQ was 150. For the first time, eugenicists had a tool that could spell out, in black and white, a person's value to society.

World War I saw widespread adoption of intelligence testing by the United States Army. The intent was that the most intelligent recruits would be sent to officer training, the least intelligent would be rejected from service, and those in the middle assigned to technical, combat, or other duties according to their scores. But the process didn't go as smoothly as its proponents hoped. Different testing methodologies were tried, there were inadequate resources for testing such large numbers of men, and many of the results were controversial.

What arose from this was a thorough revision of the scoring, developed by David Wechsler, the chief psychologist at Bellevue Psychiatric Hospital. As a young man he'd worked with the Army during its troubled attempt at implementing intelligence testing. His innovation was to grade the tests on a curve, with your score representing your placement within the distribution of all the aggregated scores. This is now the universal standard. The scoring is designed in such a way that graphing all the scores of a given population will result in a perfect bell curve. The intent is for the peak of the curve to hit exactly at a score of 100 (which should represent about 2.7% of the population), with the long tails of the curve petering out at about 50 and 150. For those of a statistical mindset, the distribution is intended to have a standard deviation of 15. Whenever the tests are revised (we're now using Stanford-Binet 5), the scoring system is reset so that the average is again 100 and the standard deviation is again 15. We still call it the IQ, even though it's no longer a quotient.

Well, we're not practicing institutionalized eugenics anymore, and IQ scores no longer restrict where we can go and what job we can have, so is all the controversy gone from IQ testing? Not hardly. It was gone, for the most part, until the 1994 publication of *The Bell Curve: Intelligence and Class Structure in American Life,* a book by Harvard experimental psychologist Richard Herrnstein and conservative political scientist Charles Murray. The controversy came raging back with a vengeance. *The Bell Curve's* central thesis pointed out many inconvenient and politically incorrect sociopolitical implications of IQ scores.

The nice way of summarizing it is that intelligence is the strongest predictor of factors such as professional success, criminal activity, and divorce rates, and thus correlates strongly with various sociopolitical and ethnic groups across the country. The harsh way of summarizing its most controversial chapters is that blacks are less intelligent than whites. This finding triggered a tsunami of academic and popular criticism that publisher Free Press rode all the way to the bank, and that kept *The Bell Curve* squarely on the best-seller list.

The most troubling finding by the authors was that intelligence appeared to be the result of a combination of both nature and nurture. In simplified terms, this means that race plays at least some role in determining intelligence. The criticism of this claim came from many different directions: That Herrnstein and Murray had used flawed weighting in their statistical measurements; that their studies were improperly controlled; that they'd ignored contradictory research; and that they'd based their research on unproven assumptions. Unfortunately it's nearly hopeless for a layperson to try and evaluate either the claims or the criticism; one quickly discovers that the statistics involved are extremely complex.

About a year after *The Bell Curve* was published and the charges of racism had been thoroughly aired, the American Psychological Association decided to write its own report to specifically address the book's findings. A diverse task force of American psychology professors was assembled to "identify, examine and summarize relevant research on intelligence." Of the difference between blacks and whites, the APA confirmed that there has long been about a 15-point difference, which is one standard deviation; but it also found that there is no clear reason for this, and there is certainly not sufficient evidence to point to a genetic cause. Society is very complicated, and many factors appear to affect intelligence. Some of these suspected factors, most of which are unproven, include nutrition, education, English skills, experience with testing, and heritability.

The APA's final conclusion was critical not just of *The Bell Curve*, but of the debate in general:

> *In a field where so many issues are unresolved and so many questions unanswered, the confident tone that has characterized most of the debate on these topics is clearly out of place. The study of intelligence does not need politicized assertions and recriminations; it needs self-restraint, reflection, and a great deal more research. The questions that remain are socially as well as scientifically important. There is no reason to think them unanswerable, but finding the answers will require a shared and sustained effort as well as the commitment of substantial scientific resources. Just such a commitment is what we strongly recommend.*

Two of these unanswered questions stand out as particularly intriguing: The racial differences, and something called the Flynn effect, and it may turn out that they're related. New Zealand political scientist Jim Flynn first noted that every time intelligence tests have been revised, average scores worldwide have gone way up, by about a standard deviation; and it's been necessary to reset 100 to a higher and higher point. People have been getting more intelligent ever since testing began, and some believe this improvement is accelerating. The reasons for the Flynn effect are unknown, but hypotheses usually center around the nurture factors for intelligence such as an increasingly intensive academic environment and healthcare. The Flynn effect is proven to change scores by at least as much as the racial differences that have been found, and it's possible (though far from evidenced) that unequal distribution of the same intelligence nurturing resources responsible for the Flynn effect may be responsible for the racial differences.

And so, while the roots of IQ testing came from the inherently negative process of identifying and culling out the worst of humanity, its future may prove to be crucial in helping everyone develop to a higher potential. Eugenics is one of those shameful follies that can't be uninvented, but its lessons may not have been entirely without fruit. When Binet and Simon first set out to learn how to find the schoolchildren who needed special help, they may have been onto something with far broader application. Theirs was not the spirit of culling, but the spirit of helping; and intelligence testing will always be linked to both.

REFERENCES & FURTHER READING

Flynn, J. "Massive IQ gains in 14 nations: What IQ tests really measure." *Psychological Bulletin.* 1 Mar. 1987, Volume 101, Number 2: 171-191.

Gottfredson, L. *Scientific American Book of the Brain.* New York: Lyons Press, 1999. 57-68.

Gould, S. *The Mismeasure of Man.* New York: Norton, 1981.

Herrnstein, R., Murray, C. *The Bell Curve: Intelligence and Class Structure in American Life.* New York: Free Press, 1994.

Kühl, S. *The Nazi Connection: Eugenics, American Racism, and German National Socialism.* New York: Oxford University Press, 1994.

Neisser, U. "Intelligence: Knowns and Unknowns." *American Psychologist.* 1 Feb. 1996, Volume 51, Number 2: 77-101.

45. Whales and Sonar

Navy sonar is said to be lethal to whales. What does the latest research actually tell us?

In this chapter, we're going to sink into the deep ocean waters, the realm of great natural denizens such as beaked whales, and of the hulking artificial beasts called submarines. In this world of darkness and near-zero visibility, audio shimmers across the frequency spectrum, from the high-frequency chatter of dolphins to the great long distance low-frequency calls of

whales. And, every once in a while, a tremendous electronic burst rips through the environment: a sonar ping. What happens next is debated worldwide. Some say the whales are driven mad, others say they are disoriented and beach themselves, some claim they are deafened or even killed outright. What's the truth? What is the real impact of Navy sonar on marine life?

Briefly, there are two basic kinds of sonar, active and passive. Active sonar sends out loud pings, much like radar, and the returning sound waves paint a picture of the surrounding environment including the location of enemy vessels. Passive sonar listens only without making any sound of its own. Active sonar is the type we're interested in.

There are different types of active sonar, primarily used in anti-submarine warfare (ASW). Low-frequency active sonar (LFAS), which operates around 300 Hz, was first suspected of causing harm to marine life, but upon further study, it was found there was no statistically significant correlation between the use of LFAS and whale strandings. Later we discovered that mid-frequency active sonar (MFAS), operating between 3000 and 4000 Hz, did in fact have such a correlation. So the specific culprit that we're looking at is mid-frequency active anti-submarine sonar.

The species that seem to be susceptible, according to observations, are pretty specific as well. They are the beaked whales, *Ziphiidae*. The danger to them is in the form of mass strandings. The definition of a mass stranding is pretty generous; two or more whales, within six days, within 74 kilometers, constitute a mass stranding. Whale strandings overall are rare enough that even a case of as few as two whales so far apart likely constitutes a related event. Whales need not die to be counted; often many of the whales are refloated and return to safety, but even these are considered to be part of the stranding event.

Yet, contrary to observations, some activists charge that much broader dangers exist. A 2010 article from the Environmental Protection Information Center, a California non-profit environmental watchdog group, states:

> *Sonar is extremely dangerous to marine mammals such as whales, dolphins and porpoises. These animals rely on their own sonar for food, navigation, mating and when high frequency sonar like the Navy is proposing to use reaches these mammals, they can be severly [sic] affected.*

High frequency sonar has not been observed to have any effect on marine mammals at all, but it's also relatively new. High frequencies are absorbed very quickly by water, so its range is much shorter; thus, it hasn't been very useful for submarines. But with more recent software, it's now being used for high resolution imaging of ice and seafloors, as well as for close range detection of small objects like mines. Toothed whales, including dolphins and porpoises, use immediate-vicinity echolocation in the same way, but at frequencies much higher than the Navy, in fact much higher than the range of human hearing

(40 kHz to 150 kHz). Social communication uses lower frequencies, because it carries much better in water.

The limited range of high frequency sonar, and the related fact that most marine life (including non-toothed whales) can't hear it at all, makes it much less likely to have any significant effect on mammals. It may, there's just no evidence for it yet; and acoustic science makes it unlikely.

The Environmental Protection Information Center's article continues:

> *Sonar has a huge impact on marine life anything from frying fish eggs, disorienting marine mammals causing them to be stranded, to permanently damaging their ears.*

"Frying fish eggs" is just silly, and it's a little bizarre that they included such a statement. Other claims have said that whales are driven insane by the noise, which is an unfounded, unprecedented, and uninformed supposition. But the strandings and hearing damage are possibilities; so let's take a look at what we've learned from the latest research.

Scientists and the Navy agree that mid frequency sonar can cause whale strandings, but so far nobody's been able to determine why, and we're just beginning to understand in what circumstances. One hypothesis has been that the noise provokes them to suddenly surface, causing decompression sickness. This remains just a guess, since so far, none of the whale carcasses examined following sonar-associated strandings has exhibited any signs of decompression sickness. So if it is occurring, it's not happening anywhere we've ever observed. It could happen way out in the open ocean where we'd never find the carcasses; we just don't know.

All the other hypotheses we have are equally elusive. These include stress, hearing loss, and disruption of the whales' feeding. Dr. Darlene Ketten is a senior scientist at the Woods Hole Oceanographic Institution who studies sonar and whale strandings. She uses CT scanning on whales and biophysical models of hearing in marine animals, and has yet to find any evidence that whales' hearing has been damaged by sonar. In every case where a stranding was associated with sonar use, Dr. Ketten found that the only discernible cause of death was the injuries sustained from the beaching itself. The whales were well fed, their hearing was intact, and they had no signs of decompression sickness. There's been no evidence to support any hypothesis. The only problem anyone has been able to find has been the big one: The whales were dead.

The best clue yet lies in underwater geography. One of the most active places for the US Navy to use sonar in exercises is off the coast of southern California, where beaked whales abound. Yet, strangely, no cases of whales stranding or otherwise being affected associated with sonar activity off California has ever been discovered. In addition, such sonar is now used worldwide, not just by many nations' navies, but also by commercial vessels, and also other types of underwater noisemaking gear such as the towed airgun arrays used by oil companies searching for new oil fields. If sonar and other manmade sounds were always harmful to nearby marine mammals, we would expect to see a far larger number of such injuries all around the world. But we don't; when and where it happens seems to be highly specific.

The latest and most comprehensive publication on this is from 2009, in the journal *Aquatic Mammals*, entitled *Correlating Military Sonar Use with Beaked Whale Mass Strandings: What Do the Historical Data Show?* The paper includes graphical timeline representations from around the world of when naval sonar exercises took place, plotted alongside mass strandings in the local region. The findings to date are intriguing. There has been a correlation of sonar and strandings in the Mediterranean and the Caribbean, but so far, no correlation in California or Japan. The reason seems to be the underwater geography concerned, called the bathymetry. Bathymetric surveys of the areas reveal that steep dropoffs close to shore, especially when confined areas exist, are the places where beaked whales are at risk from sonar. Off the coasts of Japan and California, there are relatively broad shelves, with deeper trenches being farther offshore. Beaked whales do not appear to be at risk in areas with this type of bathymetry.

The research is quite clear that we don't yet have enough answers to fully understand the reasons for this. It may be as simple as how close to shore the whales' habitats are. Most mass strandings associated with sonar have happened where the shoreline is less than 80 kilometers away from water greater than 1000 meters deep. If they're spooked by the sonar, they may simply have less free space in which to avoid it. And of these stranding events, most of them happen on specific coasts where six or more such strandings have been recorded. The evidence strongly suggests that beaked whales are at the most risk from sonar-associated strandings only on coastlines with specific bathymetric characteristics. No evidence yet supports any danger to these animals near other coastlines, such as California, or in the open ocean. That's not to claim the danger doesn't exist; merely that we don't have evidence that it does.

Navies continue to develop better techniques for managing this. Sonars are turned off when whales are detected nearby. Aerial surveys are also used to spot whales in advance, but this relies on the luck of happening to spot them near the surface. Scientists from the Woods Hole Oceanographic Institution funded by the Navy have attached non-invasive acoustic recording tags to beaked whales to learn the audio profile of their lifestyle and environment. This data helps us understand what kinds of sounds the whales make use of, and also provides better signatures for ships and submarines to listen for to help detect their presence. This research is still in its early stages, and it's work such as this that needs to be done before we'll have a complete understanding of the true risk factors.

Perhaps the most frustrating aspect of this question is that so much energy and money is spent on protests and lawsuits when neither side yet has sufficient science to support their position. Rather than spend this money learning the facts, both sides are spending money trying to enforce policy based on supposition. According to Dr. Peter Tyack at Woods Hole:

> *When the courts and the public do not get an accurate picture of the threats posed by different human activities to marine mammals and other wildlife, it distorts conservation priorities and does not serve the interests of the animals.*

There's one big shocker that brings the question of whales and sonar into proper perspective. Dr. Barbara Taylor, a cetacean specialist at NOAA's Southwest Fisheries Science Center found that a total of about 200 whales have been killed in strandings associated with naval sonar in the past 40 years. That's about five per year. It's five too many; but it's still five. Many more than that are temporarily stranded or otherwise affected, but with no lasting effects. By stark contrast, 300,000 whales and dolphins are killed every year by fishing operations. That's *sixty thousand times as many.*

Meanwhile, the lawsuits, protests, and petitions continue raging, with good intentions but little informed science to support the charges. It seems unlikely that the world's navies will make do without sonar, so continued research will remain critical to minimizing or eliminating the risk to marine mammals. If you really want to help the whales, hug a scientist, or otherwise support the research.

References & Further Reading

Buck, E., Calvert, K. *Active Military Sonar and Marine Mammals: Events and References.* Washington, DC: Congressional Research Service, 2008.

D'Amico, A., Gisiner, R., Ketten, D., Hammock, J., Johnson, C., Tyack, P., Mead, J. "Beaked Whale Strandings and Naval Exercises." *Aquatic Mammals.* 1 Jan. 2009, Volume 35, Number 4: 452-472.

Filadelfo, R., Mintz, J., Michlovich, E., D'Amico, A., Tyack, P., Ketten, D. "Correlating Military Sonar Use with Beaked Whale Mass Strandings: What Do the Historical Data Show?" *Aquatic Mammals.* 1 Jan. 2009, Volume 35, Number 4: 435-444.

Jamieson, A. "Navy Proposes Warfare Training Range in Pacific Northwest." *EPIC.* Environmental Protection Information Center, 17 Oct. 2010. Web. 6 Dec. 2010.
<http://www.wildcalifornia.org/blog/navywarfaretraining/>

Madin, K. "Supreme Court Weighs in on Whales and Sonar." *Oceanus.* Woods Hole Oceanographic Institution, 27 Mar. 2009. Web. 7 Dec. 2010.
<http://www.whoi.edu/oceanus/viewArticle.do?id=56252>

NOAA. "Sonar and Marine Mammals Fact Sheet." *Fisheries Service.* National Oceanic and Atmospheric Administration, 7 Oct. 2006. Web. 7 Dec. 2010.
<http://www.nmfs.noaa.gov/pr/pdfs/health/sonar_fact_sheet.pdf>

46. Hollywood Myths

A look into some of the classic Hollywood legends that you've always believed are true.

In this chapter, we're going to point the skeptical eye at legends of the cinema. We're not talking about ridiculous movie science, such as the idea that any computer in the world can control the air conditioning in any office suite anywhere in the world using dramatic full-motion animation of flying through electrical panels and HVAC ducts.

No, we're talking about the folklore of classic Hollywood, the enduring fables that gild the stars with mystique. So let us now dim the lights, raise the curtain, strike a dramatic note on the Wurlitzer, and fade in on:

Ben-Hur Myths

There are at least two lingering legends about *Ben-Hur*, the 1959 epic starring Charlton Heston. It was Hollywood's third iteration of the film, with black and white silent versions appearing in 1907 and 1925. The 1925 version rivaled the scope of the 1959 movie, with a chariot race in particular that was larger and more violent. The charioteers are said to have been offered a prize of $5,000 to win, for real, resulting in much on-screen mayhem.

This probably gave rise to the most common myth, that a stuntman was killed during the 1959 chariot race and that its footage was left in the final cut. Although a few sources have made this claim, the studio's records, and the statements of most of those involved including Charlton Heston, say that no stuntmen or horses were seriously hurt. The sequence was very carefully shot over several weeks. The worst injury came from a stuntman whose chin was smashed and cut when he unexpectedly flipped out over the front of his chariot. Articulated dummies were put to very effective use throughout this scene, and the guys you see getting trampled or run over are all dummies.

The source of the myth is probably the 1925 film. Records were poorly kept and incidents often went unreported in those days, so we don't know for sure, but most film historians agree that at least one stuntman was killed, and at least a few horses were injured badly enough that they had to be put down. A lot of this happened in one particular crash where several chariots came around a blind corner and struck another that had overturned.

The other myth from *Ben-Hur* is that a red sports car, sometimes described as a Ferrari, can be seen in the background of the 1959 chariot race. There's a good reason you won't find any screen captures of this on the Internet; it seems to be a completely untrue rumor. I watched the entire scene carefully on the DVD, and others on the Internet have gone through it frame-by-frame. No red car has been found yet; though there are numerous tire tracks visible onscreen from the camera trucks, which may be the source of the myth.

Goldfinger's Golden Girl

We all remember the classic scene where Sean Connery comes across the gold-painted corpse of Goldfinger's assistant whom he had seduced to betray her boss. Later he observes that she had died of skin suffocation, and that painting the entire body would be lethal unless you leave a small bare patch at the base of the spine, "to allow the skin to breathe". This danger subsequently became something of an urban legend.

Perhaps the writers forgot that we breathe through our lungs, not through our skin. Today, performers are routinely bodypainted, even with latex. People take mud baths. There are certainly other things that can go wrong when you coat your entire body — overheating and hypothermia are two real possibilities — but "skin suffocation" is not one of them.

The *Poltergeist* Curse

The story goes that all three children who starred in *Poltergeist* were dead soon after its release, as were a number of other actors from the series. It's partially true, but a few hits and many misses does not really make a good legend.

Dominique Dunne, who played the older daughter, was murdered at age 22 only five months after the movie was released. Heather O'Rourke, the little girl, died from a medical condition at age 12, but only after completing both sequels. In a blow to the curse, Oliver Robins, who played the son, is still alive and well, and is writing and directing.

Other actors from the original *Poltergeist* and the sequels have since passed away as well, but of expected age related conditions. It's a little like saying there's a Gettysburg Address curse, since everyone who attended is now dead.

The *Wizard of Oz* Hanging

There's a famous legend that an actor playing one of the munchkins in *The Wizard of Oz* was distraught over a lost love and decided to hang himself in the background of a live scene shot in the movie. Some versions of the story have it as a stagehand who may have fallen accidentally. Somehow nobody noticed it at the time and the shot was left in the final cut. Right after Dorothy and the Scarecrow pick up the Tin Woodsman, they link arms and go happily singing down the yellow brick road. In the center of the screen, between two trees against the blue backdrop, something can be seen falling to the ground. A few seconds later there is another large movement in the same spot. It's too far away and indistinct to make a clear judgment, but with a little imagination, I suppose it could be taken as someone being hanged.

Fortunately, there are no records of any hanged corpses being discovered when the set was struck. What did happen, though, was that various large birds were borrowed from the Los Angeles Zoo to roam freely around and make the sets look more wild. Although the movement is indistinct, it takes very little imagination to see one of those birds jumping down from the tree then spreading its wings. Or doing some other bird thing. No record survives of bird poop being cleaned up in that spot, so this is just one of those mysteries that we'll never solve. You judge which you think is more likely: A suicide where no body was found and nobody was missing; or some large birds strutting about that are known to have been there.

Brandon Lee's Death

The mysterious death of Bruce Lee's son was sure to achieve a cult status all its own. The story goes that actor Brandon Lee was shot on the set of his final film, *The Crow*, in the middle of filming a scene; and that his death was left in the final cut of the movie. Many of the same conspiracy claims surrounding his father's death surfaced again: He was assassinated for revealing martial arts secrets, killed by organized crime, or some other such thing.

Brandon Lee was indeed shot while filming. It was a tragic accident involving a gun firing blanks. A fragment of a dummy bullet, from a previous scene, was lodged in the gun and fired into Lee, fatally wounding him. Some mystery remains surrounding the film of the incident, with some saying it was destroyed, and others saying it was confiscated by police. It was not used in the movie. The scene was rewritten and reshot using a double, and the manner of his death is different than what happened in the fatal accident.

No credible evidence links the tragedy in any way to organized crime or martial arts overlords. Everything that happened was fully explained by the events of the day, no external mysterious forces required.

Astronauts, Aliens, and Ape-Men

Steven Spielberg's Covert Beginnings

One of director Steven Spielberg's most enduring pieces of fictional entertainment is the story he used to tell about how he got started in show business. He snuck off of a Universal Studios tram, slunk around the studio, and moved into a vacant office, adding his own name to the building directory. From there he met people, introducing himself as a new director on the lot, and finally weaseled his way into his first directing jobs.

Never let the truth get in the way of a good story, especially if the truth is really mundane. Spielberg was once shown around the lot by a friend of a friend of his father's, and returned the next summer for an unpaid internship working on purchase orders. He did make the most of this opportunity, though, finally landing a job directing a TV episode after several summers running errands and stacking paper. The rest is movie history. But among those who worked there with him at the time, none remembers any but legitimate work-related reasons for him to be wherever he was.

James Dean's Killer Car

James Dean, one of Hollywood's prototypical outcast bad boys, enjoyed his motor racing at least as much as he enjoyed acting. To this end he bought a 1955 Porsche 550 Spyder, in which he was infamously killed in a road accident on his way to a race. The legend says that anyone who owns the wrecked Spyder has bad luck; they die or get injured by it or something like that.

The car was pretty comprehensively destroyed in the crash, and its few useful (and valuable) mechanical components were parted out by the insurance company. When the wreck was initially purchased, a man's legs are said to have been broken when it was delivered on a trailer. It's not too surprising, since the car was a non-rolling lump and probably had to be awkwardly lifted onto dollies to roll it down the ramp. Two amateur racers installed the car's engine and transaxle in their own cars. One was later killed and the other was injured in separate racing accidents in cars using the Spyder's components, however none of the James Dean parts caused the accident. A whole great long string of horrible injuries are said to be associated with the car's body, which went on tour with the California Highway Patrol as an exhibit in the late 1950's, but there is no reliable documentation that any of these injuries actually happened.

Bits of the car disappeared (some are currently displayed in various museums), and it's said that the remaining body was mysteriously lost during transport. However no police report was ever filed regarding a theft, so the safe money says that there's probably no appreciably sized remnants of the car unaccounted for; the families of the two amateur racers still have the parts they bought. Certainly there's no good evidence that unusual accidents were associated with the car or its parts, and enough details of the legend are demonstrably fabricated to cast doubt on the whole idea.

James Dean in his Porsche 550 Spyder

The Wizard of Oz and Pink Floyd

The story goes that if you watch *The Wizard of Oz* while listening to Pink Floyd's album *Dark Side of the Moon*, there are moments where they appear to synchronize, where certain lyrics seem relevant to the action onscreen. This combination has been popularly dubbed *Dark Side of the Rainbow*.

The band members and engineer Alan Parsons all deny it. It's one of those cases where for every moment that does seem to match up, there are 100 other moments that don't; and no clear agreement among believers on what matches and what doesn't. People have claimed many such matches between various movies and albums. I'm not going to try and talk you out of this one. They're both works of art, and art is in the eye of the beholder; but if you want someone (who knows) to admit it was done deliberately, you're going to have to keep waiting.

Three Men and a Ghost

There's one myth that I almost didn't include because I was trying to stick to classic Hollywood and this was such a stupid movie and an even stupider myth, but it seems to have grown long enough legs to warrant a mention. In *Three Men and a Baby*, the story goes that the ghost of a boy is visible standing in some curtains in the background. Screen caps are all over the Internet if you want to take a look. But one of the deleted scenes shows a better view of the object, and it's a near life-size cardboard cutout of Ted Danson wearing a tux and top hat. Viewed out of focus and from across the room, the top hat looks like a young boy's mussed-up hair. If you try to look this up on the Internet, it's almost impossible to find a page that does not also give the explanation; so I doubt you can find anyone who's heard of this myth but doesn't know about the cardboard cutout. Nevertheless, Ted Danson's indomitable star power (hint of sarcasm) seems to keep this one near the top of every list of Hollywood legends.

Buried at the Magic Kingdom

This one isn't about a movie, but since Disney is a movie studio, I'll include a myth about their Magic Kingdom theme parks. The story goes that there are people buried at the parks, possibly including Walt Disney himself, said to be cryogenically preserved under Disney World.

The stories are completely untrue, in fact I couldn't find any credible evidence worthy of examination. If there is any, let me know. Walt himself was cremated, and his ashes reside at Forest Lawn in Los Angeles. But this leads us to a creepy fact that almost confirms the rumor: On many occasions, Disney security has caught guests scattering the ashes of deceased Disney fans at theme parks. Most of the evidence of this is anecdotal, as you're not likely to see Disney sending a press release to the newspapers. But once when police were called when a woman was seen doing this at the Pirates of the Caribbean ride, Disney employees began emailing bloggers and columnists who follow Disney that it happens quite frequently.

So although you won't find anyone officially buried at the Magic Kingdom, the remains of its most *enduring* fans... are everywhere.

The Superman Curse

There are various versions of the so-called Superman curse, centering around the idea that anyone who plays Superman meets an untimely end. It's also said that George Reeves, the original television Superman, went crazy and thought he could actually fly, and tossed himself from a building. That one's easily disproven: George Reeves died of a gunshot wound ruled a suicide, but like so many suicides, enough of his loved ones said that "he'd never do such a thing" that dark rumors of murder and hit men persisted, as promoted by the 2006 movie *Hollywoodland*.

Although bigscreen Superman Christopher Reeve was paralyzed and died nine years later, his career flourished in the seven years since he'd last played Superman; and when he died, he died a much beloved philanthropist and spokesman. Even George Reeves' career was doing well; the TV show had been renewed for another season when he died and was very successful.

The other Supermen also serve as evidence that playing Superman is less of a curse than a blessing. The original movie serial Superman, Kirk Alyn, was successful enough that he was typecast and chose to leave the industry, happily retiring to Arizona. Dean Cain of *Lois & Clark* is alive, well, and happy, as is Tom Welling of *Smallville*. Bud Collyer played Superman on the radio for 11 years, and in cartoons for two years, and went on to have a full and successful career.

We should all be so super lucky to be stricken with such a super curse.

John Wayne and the Nevada Test Site

Supposedly, John Wayne's death from cancer was caused by his work in the Utah desert in 1954 on the 1956 Howard Hughes film *The Conqueror*, a movie widely regarded as Wayne's worst. The location near St. George, Utah, is notorious for being downwind from the Nevada Test Site, where a large number of atomic weapons had been detonated in prior years, and thus was the recipient of much radioactive fallout. Wayne's co-stars Susan Hayward and Agnes Moorehead also died of cancer; in fact, by the time *People* magazine checked up on all 220 cast and crew for a 1980 article, 91 of them had contracted some form of cancer, and 46 had died of cancer.

People's inspiration was apparently a 1979 article in the tabloid *The Star* by Peter Brennan who merely speculated about the coincidence without doing any real research. It was repeated by such newspapers as the New York Post (August 6, 1979) and the Los Angeles Times (August 6, 1979). *People* went a step further, talking to a few experts and managing to track down the history of the cast and crew. This article was what really started the story; in fact, virtually anything you might find about this story takes its quotes directly from *People*. One of the most often borrowed was from an enthusiastic fallout activist, Dr. Robert Pendleton at the University of Utah, who said:

With these numbers, this case could qualify as an epidemic. The connection between fallout radiation and cancer in individual cases has been practically impossible to prove conclusively. But in a group this size you'd expect only 30-some cancers to develop. With 91, I think the tie-in to their exposure on the set of The Conqueror would hold up even in a court of law.

But it didn't, at least not for residents of St. George, Utah, often referred to as the "downwinders", when attorneys went door-to-door in the 1970's. *The Times* of London reported that some 700 such lawsuits were unsuccessful. However, ten years after the *People* magazine article, the Radiation Exposure Compensation Act was passed and has since paid out over $1.5 billion, including many payments to people who had only to prove that they lived in certain counties during a certain time period, and had one of a list of approved diseases. Although this makes it sound like the link must have been proven, science doesn't depend on what politicians were able to convince bureaucrats to do.

And what science has found, contrary to what's reported in virtually every article published on the subject, is that any link between the film crew's cancers and the atomic tests is far from confirmed. First of all, the numbers reported by *People* are right in the range of what we might expect to find in a random sample. According to the National Cancer Institute, in 1980 the chances of being diagnosed with a cancer sometime in your lifetime was about 41%, with mortality at 21.7%. And, right on the button, *People's* survey of *The Conqueror's* crew found a 41.4% incidence with 20.7% mortality. (These numbers make an assumption of an age group of 20-55 at the time of filming.)

A 1979 study in the *New England Journal of Medicine* found no consistent pattern of correlation between childhood cancers and fallout exposure in the Utah counties, with the exception of leukemia. For reasons unknown, leukemia rates were about half that of the United States at large, but after the fallout period, this increased to just slightly above the normal rate. The authors were unable to correlate either leukemia or other cancers to fallout. Considering that the film crew spent only a few weeks there, instead of their whole lives like the people who were studied, it seems highly unlikely that they were affected.

But we can't make that declaration for certain. The data we have for the film crew is totally inadequate. Most crucial factors are unknown, like age, age of incidence, types of cancer, heredity, dose-response, and other risk factors each may have had — like John Wayne's smoking of five packs a day. And, of course, "cancer" is not one disease; it is hundreds of different diseases. Plus there's an obvious alternate explanation: The cast and crew simply got old in those intervening decades.

What about Dr. Pendleton's gloomy remarks? In an email to researcher Dylan Jim Esson, a colleague of Pendleton's, Lynn Anspaugh, said that Pendleton's reported comments were uncharacteristic and she thought they were more likely the result of media sensationalism. According to her analysis of the fallout readings from the time and place of *The Conqueror's* filming, she calculated that the crew received no more than 1 to 4 millirems of radiation, which was less than normal background levels. Pendleton himself had recorded high levels of radiation only when a fallout cloud was directly overhead the day following a test, and normal at other times. The most recent tests had been more than a year prior to the filming, so Anspaugh's calculations are not surprising.

From all the data we have, it was perfectly safe for the film crew, and their reported cancer histories show no unusual ill effects.

So there we have it, another line of evidence that Hollywood myths are all just a part of the show. But if it were not for these romantic tales, Hollywood would not be Hollywood. Whenever you hear a story that seems too incredible to be true, *don't* be skeptical if it's from Hollywood. Instead suspend your disbelief, and enjoy the show.

References & Further Reading

Beath, W. *The Death of James Dean.* New York: Grove Press, 1986. 9-10.

Esson, D.J. "Did 'Dirty Harry' Kill John Wayne? Media Sensationalism and the Filming of 'The Conqueror' in the Wake of Atomic Testing." *Utah Historical Quarterly.* 1 Jun. 2003, Volume 71, Issue 3: 250-265.

Freiman, R., Wallace, L. *The Story of the Making of Ben-Hur.* New York: Random House, 1959.

Harmetz, A. *The Making of The Wizard of Oz.* New York: Dell Publishing, 1989.

Jackovich, K., Sennet, M. "The Children of John Wayne, Susan Hayward and Dick Powell Fear That Fallout Killed Their Parents." *People.* 10 Nov. 1980, Volume 14, Number 19: 42.

Lyon, J., Klauber, M., Gardner, J., Udall, K. "Childhood leukemias associated with fallout from nuclear testing." *New England Journal of Medicine.* 22 Feb. 1979, Volume 300, Number 8: 397-402.

Meehan, P. *Cinema of the Psychic Realm: A Critical Survey.* Jefferson: McFarland, 2009. 98.

Mikkelson, B. "Movies." *Urban Legends Reference Pages.* Snopes.com, 1 Jan. 2010. Web. 12 Dec. 2010. <http://snopes.com/movies/>

Shaw, M. "Was The Movie The Conqueror Really Cursed? A Look At Radiation Paranoia." *Health News Digest.* Interscan Corporation, 14 Sep. 2009. Web. 25 Dec. 2010. <http://www.gasdetection.com/news2/health_news_digest225.html>

Yoshino, K. "A ride to the great beyond at Disneyland." *Los Angeles Times.* 14 Nov. 2007, Newspaper.

47. Gluten Free Diets

Are gluten free diets really good for your general wellness?

In this chapter, we're going to point the skeptical eye at the promoters of gluten free diets. Gluten comes from wheat and at some level, just about any commercially available food either contains wheat or has trace contamination from wheat, so a gluten free diet is much easier said than done. It's become one of the new fads in health food stores, and some claim that such a diet can treat autism or obesity or any of a wide variety of conditions. Is gluten really something that would be good for most people to avoid? What exactly is it, what's it used for, and how does it affect our bodies?

The history of human culture is closely tied to the history of bread. Bread was one of our earliest portable foods, which made it possible to take long journeys. Its carbohydrate content made it a high-energy food, and combined with its light weight, bread was about the best food you could have with you. Bread made it possible for humans to migrate, for armies to march, and for history to be made.

The earliest breads made from crushed corn or plant roots were poor in quality; they were like crumbling wafers that were hard to carry or preserve. Paradoxically, it was the development of agriculture that both kept people in place and allowed them to move. Wheats and other grains began to be used for bread; and as it turned out, wheats and a few related cereals like rye and barley contain a protein called gluten. Gluten is a long, tough molecule, and it's what gives modern bread dough its sponginess and elasticity. Bread baked from wheat flour resulted in loaves that didn't fall apart, and could be transported great distances. Gluten built the bread that built the world.

And since then, gluten has been used in a good many other foods as well. It's handy as a protein supplement, and as an all-natural way to add elasticity to foods. Such products as ketchup and ice cream are commonly thickened with gluten. Some pet foods use gluten as a way to boost the protein content without adding meat. Almost all imitation meats and cheeses prized by vegetarians are based on wheat gluten. And gluten is not just limited to food. Its long, tough molecules make it a key ingredient in some new bio-plastic materials as an alternative to petrochemicals. Gluten is even commonly used in cosmetics such as lipstick to help firm it up.

Wheat

But there's been a growing trend in recent years to view gluten in a negative light. It is true that a small number of people are born with gluten sensitivities that reduce their ability to tolerate it to varying degrees. Something of a non-sequitur line of reasoning has followed, that if some people can't tolerate it, it therefore must be generally bad for everyone. Gluten's increasingly ubiquitous application in a growing number of food products has triggered suspicion of the food industry's motives. As a result, some promoters of fad diets and various health schemes are now advocating gluten free diets.

Gluten free diets actually are necessary for some people, and advisable for others. Without going into too much detail, the gluten protein consists of two other proteins, a prolamin and a glutelin. The principal prolamins and glutelins in wheat are gliadin and glutenin. Generally, when we discuss gluten sensitivities, gliadin and glutenin are the specific culprits. So let's take a quick look at the three basic types of gluten sensitivity. These are all legitimate medical conditions. They're quite rare, but they are real and patients need to be aware.

The first is celiac disease (CD), or gluten-sensitive enteropathy. This is an autoimmune disease of the small intestine that occurs in people with a genetic predisposition. It's not caused by gluten and you can't develop it by eating gluten, but if you're one of the unlucky few born with the gene, and you develop CD (which not everyone does), eating gluten will cause an adverse reaction. The immune system inside the bowel tissue improperly reacts to the gliadin protein, which causes inflammation of the bowel tissue, and interferes with your body's ability to absorb nutrients from food. There's no cure for CD, and the only way to live with it is to adopt a gluten free diet for the rest of your life. Somewhere between one and eight tenths of one percent of Americans have this (1-8 in 1,000), give or take; the number is not well known.

A wheat allergy is very different, and can be harder to track down since there are many different components of wheats and other grains that it's possible to be allergic to. A wheat allergy is not a single condition; it is any of a great number of possible allergies. The symptoms are similar to what we expect from most allergies: hay fever type symptoms, hives, asthma, and swelling. More serious effects in the worst cases can include anaphylaxis, palpitations, swollen throat, diarrhea, even arthritis. Unlike CD patients, sufferers of wheat allergies need not necessarily avoid all wheat products. The allergy is usually pretty specific and only some foods may need to be avoided. Standard allergy treatment with any of a variety of drugs such as histamine blockers or leukotriene antagonists may prove effective enough to allow the patient to live with a normal diet. You need not eat wheat to have an allergic reaction; many workers who contact wheat can experience allergies as well. It's very difficult to attach a number to how many people have some level of allergy to some type of wheat related protein, but it's probably somewhere in the single digit percentage points.

There's also a third type of gluten sensitivity, and that's gluten sensitive idiopathic neuropathy. Idiopathic means the exact cause is not known, and a neuropathy is a disease of the nerves. Symptoms can include numbness or tingling in the extremities, or problems with muscular coordination often evidenced when walking, or even spasticity resembling epilepsy. Diagnosing this neuropathy has been really problematic. First, a common blood test for anti-gliadin antibodies frequently produces false positives, since many people have this antibody. And sometimes, sufferers may actually have a subclinical celiac disease instead. (Subclinical means it doesn't yet show up on tests or symptomatically.) Good numbers are not known on how many people may have a gluten sensitivity neuropathy, but it's probably in the range of a small fraction of 1%.

Yet those whose business is the sale of gluten free products would often have us believe that many more of us should buy them. GlutenFree.com and GlutenFreeMall.com claim their products help people with autism or ADHD, which is completely untrue according to all the science we have. The autism claim in particular is broadly repeated across the autism activist community. The treatment of autism with a gluten free diet has been studied a number of times with varying results, but so far no well designed studies have shown any plausible benefit. A 2006 double-blinded study published in the *Journal of Autism and Developmental Disorders* tested children with and without autism, on gluten-free and placebo controlled diets, and found no significant differences in any group.

Do an Internet search for "gluten free" and you'll find the term being misused by sellers of organic foods and other products, even vegan products and things sold as "cruelty free". Gluten is a purely vegetable, vegan substance that is, in every way, organic and all natural. So in many of these cases, the marketing boast "gluten free" exactly contradicts the vendor's claim of being vegan friendly. If you're a vegan, products containing gluten should be at the top of your list. It's an all-natural wheat protein.

Many naturopaths routinely list gluten as a potential cause of disease in general. This is a medically bizarre claim. Proteins are essential for nutrition, and there is no evidence that incidence of disease increased worldwide once wheat grain became a staple. It's true that bread itself is a rich source of carbohydrates, which are not essential and can be safely minimized in the diet, but this is true of gluten-free breads as well. By no logic should the strategy of avoiding carbohydrates be misconstrued as avoiding gluten.

So even if gluten is not the cause of any specific disease, at least for the vast majority of us who were not born with a gluten sensitivity, might it not be wise to still leave it out of our diets anyway, better safe than sorry? Keep in mind that a gluten free diet is no trivial matter. Every meal needs to be rethought, and many ingredients you always considered basic kitchen necessities will have to be thrown out. Forget most restaurant meals. Forget most alcoholic beverages, and even many products labeled gluten-free, as many of these continue to be found to be contaminated with gluten-containing cereals.

The belief that a gluten-free diet is a good idea anyway has also been studied, and so far the only groups we've found that it may actually be somewhat helpful for are patients with Parkinson's disease, multiple sclerosis, and a few other conditions. As far as general wellness goes, there's neither a sound theory nor any evidence. The vast majority of people currently avoiding gluten for presumed health benefits are doing so for no nutritionally plausible reason. Gluten is not a fat or a carbohydrate that you might reasonably want to avoid; it's a protein that your body uses.

So think of gluten sensitivities in the same way you'd think of bee stings or peanut allergies: of great and very real concern to a small number of people, of some concern for a few more, and of no concern to the rest of us. Don't let anyone tell you that gluten is harming you in some way that's so far not supported by any science, or that you should avoid it for the purpose of general wellness. For most of us, gluten is our friend; but never forget that it is also, like many compounds, definitely harmful to some.

REFERENCES & FURTHER READING

Berne, A. "The Accidental Vegetarian: Chefs have no beef with mock meat." *San Francisco Chronicle.* 19 Sep. 2007, Newspaper.

Elder, J.H., Shankar, M., Shuster, J., Theriaque, D., Burns, S., Sherrill, L. "The gluten-free, casein-free diet in autism: results of a preliminary double blind clinical trial." *Journal of Autism and Developmental Disorders.* 1 Mar. 2006, Volume 36, Number 3: 413-420.

Herbert, J., Sharp, I., Gaudiano, B. "Separating Fact from Fiction in the Etiology and Treatment of Autism: A Scientific Review of the Evidence." *Quackwatch.* Stephen Barrett, M.D., 13 Jun. 2003. Web. 31 Dec. 2010. <http://www.quackwatch.org/01QuackeryRelatedTopics/autism.html>

IWGA. "Wheat Gluten Applications." *International Wheat Gluten Association.* International Wheat Gluten Association, 16 Jul. 2004. Web. 31 Dec. 2010. <http://www.iwga.net/04_pet.htm>

Layton, L. "3 years after deadline, FDA still hasn't defined 'gluten-free'." *Washington Post.* 28 Apr. 2011, Newspaper.

Rewers, M. "Epidemiology of Celiac Disease: What Are the Prevalence, Incidence, and Progression of Celiac Disease?" *Gastroenterology.* 1 Jan. 2005, Volume 128, Number 4, Supplement 1: 47-51.

48. Mystery Spots

Are the various "mystery spots" around the world actually gravitational anomalies, or might the explanation lie elsewhere?

My original introduction to the skeptic community came through an unlikely channel. One day my friend John and I were exploring an old abandoned mine in the Mojave Desert called the Gunsight Mine. While deep inside at one point, eating our lunch in a gallery filled with square-set timber bracing, we were struck by the impulse to make an impromptu video. I dropped a handful of gravel and it fell up to the ceiling. I called it a "gravitational anomaly" and put it up on YouTube. Now, it was an obvious joke, and neither John nor I had any inkling that anybody might take it seriously or think it was real. But take it seriously they did. Smart people began emailing me and asking if there was some magnetic phenomenon or a strong wind in the mine. Someone sent it to Dr. Phil Plait and he posted it on his popular Bad Astronomy blog, and that's how I originally met Phil. Most people immediately see the gimmick and probably get a kick out of it, some have their minds blown by the apparent phenomenon, and judging by the comments, some derisively think it was a deliberate attempt to deceive. But everyone has *some* reaction to it: wonder, anger, or a good laugh.

Without knowing it, I had stumbled upon the reason for the success of the many so-called "mystery spots" around the world. Some of these apparent gravitational anomalies occur naturally, and some of them are purpose-built as attractions, but everyone either loves them or hates them. Let's point the skeptical eye at what mystery spots are, how they work, and most importantly, *why* they work. What is it about our brains that wants to interpret things wrong?

Typically, commercial mystery spot attractions cobble together a fictional account of how their location was "discovered". The Saint Ignace Mystery Spot in Michigan says:

> In the early 1950's, 3 surveyors... stumbled across an area of land where their surveying equipment didn't seem to work properly. No matter how many times they tried to level their tripod, the plum-bob would always be drawn far to the east, even as the level was reading level.

The Oregon Vortex claims:

> The Oregon Vortex, goes way back to the time of the Native Americans. Their horses would not come into the affected area, so they wouldn't. The Native Americans called the area the "Forbidden Ground", a place to be shunned.

The Mystery Spot in Santa Cruz, California says:

> ...We noted the compass to vary a small amount on the transit... There was no barbed wire fence near where we were at the time, and as far as we knew, no excessive mineral in this ground... We felt very light headed or top heavy, felt like something trying to force us right off the hill.

Like many similar attractions around the world, these mystery spots consist of crazily built wooden cabins built on sharp angles; the range of 18 to 25 degrees off of level is common. Within these structures, any number of conventional optical illusions are constructed and can be performed. A person appears taller who stands under the low end of a tilted beam than a person who stands under the high end. Balls will seem to roll and water will seem to flow uphill. A person can sit in a chair halfway up the wall supported only by its back legs.

It's worth pointing out that there is no such thing as an actual gravity spot, a place where gravity seems to work sideways or otherwise unexpectedly; at least, not that a person would ever be able to detect. The whole Earth is, however, scattered with gravity anomalies. Gravity maps show the Earth's surface in color, where the various colors represent the difference between the measured gravity and the gravity predicted from a theoretical reference geoid. We can make these measurements from space using satellites such as ESA's Gravity field and steady-state Ocean Circulation Explorer (GOCE), and they can also be made on the ground using accelerometers that are essentially very sensitive scales. The reason the Earth's gravity is not constant is that the density of the Earth's interior varies with different minerals. We use gravity maps for a great variety of applications: Looking for oil, salt deposits, potential volcanic activity, even mapping ocean currents.

We measure gravity in Gals. A Gal, named for Galileo Galilei, is a unit of acceleration used in gravimetry, and is equivalent to 1 centimeter per second squared ($1cm/s^2$). Worldwide, standard gravity at sea level averages out to about 980 Gals. This means that the force you feel pulling you into the ground is equivalent to what you'd feel in space if your spaceship was accelerating at 980 centimeters per second per second. If you go higher, further from the Earth's mass, gravity decreases; and gravity at the top of Mt. Everest is about 2 Gals less. So what about these variations that we measure? Their maximum range is a few hundred milligals and usually a lot less, way less than normal topographical variance. So if a person is unable to feel the difference in gravity between the ski lodge at the base and the top of a ski lift, you're certainly unable to detect the much more subtle variations that the Earth's maximum natural fluctuations can throw at you.

Given this context, it's clear that the gimmicks seen in mystery spots, like a ball rolling uphill or a person standing sideways on a wall, would require a gravity change many orders of magnitude greater than variances on Earth might ever hope to account for. If gravity was actually operating sideways in these parks, there would have to be a dense mass with gravity approximately equivalent to that of the Earth off to one side, in which case it would be impossible to hide this effect outside the bounds of the park. Whatever's going on inside mystery spots can be conclusively dismissed as having anything to do with gravity. Therefore, the explanation must lie elsewhere.

Like all optical illusions, mystery spot tricks work because our brain tries to improperly apply the rules it has learned about the structure of the world. Objects that are farther away appear closer together and smaller. Lines that converge indicate those lines are parallel but receding into the distance. These are just a few of the rules our brains learn as we develop, and they operate at a fundamental level in our perception. So even though we know at an intellectual level that the cabin is merely tilted at an angle, our spatial brains still try to feed us an interpretation that says the water is indeed flowing uphill, and that person is indeed standing on the wall. This perceptual conflict is still enjoyable, no matter how well you know the way it works.

Two researchers from the University of California, Berkeley studied the Santa Cruz Mystery Spot and published their findings in the journal *Psychological Science* in 1999. In it, they proposed a framework called orientation framing theory to describe how the brain's visual processing is guided by spatial frames of reference. They noted a number of illusions at the Mystery Spot as examples of familiar tilt-induced illusions such as the Ponzo illusion, the Zöllner illusion, the Poggendorf and Wündt-Hering illusions, and others. No mysterious gravitational effects needed.

This theory works not only with the artificially constructed attractions, but with natural "mystery spots" as well. Gravity Hill in New Paris, Pennsylvania is one such place, and it's fairly representative of the scores of such hills that are well known. Many of them are named "Gravity Hill", "Spook Hill", "Magnetic Hill", or some similar name. Like most, the Pennsylvania Gravity Hill is a road where, if you put your car into neutral at the bottom, you will roll uphill. Throngs of curious amateur investigators frequent these spots, plumb bobs and surveying equipment in hand, to try and figure out what's going on.

Many of these locations have a few things in common. They're often nestled among rolling hills that obscure the horizon, and are away from tall buildings or other structures that may provide a reference plane for the horizontal and vertical. Tree trunks and other natural features are often the only available references, leaving it easy to misjudge the slope of a road that's among hills that are higher at one end of the road than the other. It's the classical optical illusion, again explained by orientation framing theory.

Of course, plumb bobs and surveying equipment have their weaknesses. If you think gravity might be acting wrong, a plumb bob is not a reliable tool. Surveying equipment is fine and dandy, but the average layperson doesn't have access to it or the knowledge and resources to use it properly. But one tool that everyone can get their hands on is a GPS, which operates independently of gravity or magnetism or anything else you think might be going on. Take an altitude reading at each end of a gravity hill, and you'll find that what your eyes told you was the high end was actually the low end. Just keep in mind that GPS units have error ranges, and you'll want to use two or three different units and let them sit for a good long time at each point to get a good reading.

Everyone also has access to good quality radar elevation mapping of the whole planet: Google Earth. And with a little bit of research and a few dollars, you can get a printed topographical map. So far, none of the known Gravity Hills has been found to actually roll objects uphill, once a proper elevation survey is done. The Pennsylvania Gravity Hill has been subjected to many such examinations. One group, called the Enigma Project, published their results online. Their GPS survey found that over the 416-foot length of the road marked off with "Start" and "End" painted on the road, it dropped 13 feet, even though to the eye it appeared to rise. They then confirmed these findings using a tripod mounted surveying laser. Their measurements agree with the commercially available topographic map of the area, and with the elevations found on Google Earth.

It's fun to experience an illusion, and really mind blowing to wonder about it when you don't know the cause, whether it's ghosts pushing your car or a gravitational anomaly lifting you up onto the wall. But for me, the funnest part of mystery spots is comprehending the illusion, and matching what you see to what you know... especially when your brain says you can't.

REFERENCES & FURTHER READING

Eagleman, D. "Visual Illusions and Neurobiology." *Nature Reviews Neuroscience.* 1 Jan. 2001, Volume 2, Number 12: 920-926.

Editors. "Mystery Spots Explained." *Sandlot Science.* SandlotScience.com, 25 Oct. 2005. Web. 7 Jan. 2011.
<http://www.sandlotscience.com/MysterySpots/Mystery_Spots_1.htm>

ESA. "ESA's Gravity Mission GOCE." *Living Planet Programme.* European Space Agency, 16 Sep. 2010. Web. 7 Jan. 2011.
<http://www.esa.int/esaLP/LPgoce.html>

Frizzell, M. "An Examination of the Bedford County Gravity Hill." *The Enigma Project: A Documentation of the Paranormal.* M. A. Frizzell, 30 Mar. 2008. Web. 5 Jan. 2011.
<http://userpages.umbc.edu/~frizzell/pabedcogravhill.html>

Gregory, R. "Knowledge in perception and illusion." *Philosophical Transactions of the Royal Society of London.* 1 Jan. 1997, Volume 352: 1121-1128.

Shimamura, A., Prinzmetal, W. "The Mystery Spot Illusion and Its Relation to Other Visual Illusions." *Psychological Science.* 1 Nov. 1999, Volume 10, Number 6: 501-507.

Sieveking, P. "Gravity Anomalies." *The Fortean Times.* 1 Dec. 2003, December 2003.

49. THE ALIEN BURIED IN TEXAS

Some believe a space alien is buried in a rural cemetery in Aurora, Texas.

You can drive right through the hamlet of Aurora, Texas in only one minute, and never know anything ever happened there. Highway 114 winds through the low rolling country, carrying trucks and people between more important cities, passing right by the hardly a dozen storefronts. Twelve hundred quiet residents and some lazy insects mostly watch the rest of the world go by, and things are pretty slow and relaxed here. This is a small story from a small town, one you've probably never heard; and at the same time, it just might be the biggest story you ever will.

It's not the place you'd expect television crews to descend upon, but so they have, and in force. In 2005, an episode of The History Channel's *UFO Files* was shot here, as was a 2008 episode of its spinoff *UFO Hunters*. These television shows dramatized Aurora's one claim to fame: That on April 17, 1897, an alien spacecraft crashed, killing one occupant, who was then buried in Aurora's small, quiet cemetery. Today, the grave remains, undisturbed and unopened below the bent limb of an overhanging tree, purportedly containing the remains of an alien visitor.

The story was reported in the newspaper in 1897, and many decades later, modern UFOlogists took a big interest in it. In 1973, some of the biggest names in UFOlogy spent months in Aurora checking up on the old story. The were led by Bill Case and Earl Watts of MUFON (the Mutual UFO Network, then known as the Midwest UFO Network), Hayden Hewes and Tommy Blann of the International UFO Bureau, and Fred Kelley, a professional treasure hunter. MUFON produced a 199-page report you can download containing the history, newspaper reports, metallurgy reports, interviews with Aurora residents, and just about anything else you can think of. The MUFON UFOlogists have concluded that the evidence indicates the story is true as told in the newspaper.

It sounds like one of those mysteries that should be the world's easiest to solve: Simply exhume the grave and see what we find. But it's not as simple as that. The existence of the grave is known only by metal detector readings once taken by Earl Watts. Near where the readings are believed to have been found was a piece of native sandstone with what's variously described as either a triangular carving or a V-shaped crack in it, believed by the UFOlogists to have been a gravestone. But, the town removed the rock to hamper any potential efforts to dig up the graveyard, and nobody's been able to find anything on a metal detector ever since.

According to the story, the airship struck the windmill of one Judge Proctor, destroying both and killing the pilot. The UFOlogists combed the property looking for evidence of the century-old crash, and they found many bits of assorted types of metal scattered in such a way that they believe they must have been driven forcefully into their current locations by an explosion. The MUFON report acknowledges the obvious difficulty with this conclusion: that people had been living and working here for over 100 years and one could only expect to find junk and debris everywhere. (During the search, a boy even found a half dollar from 1856.) They finally focused on a single specimen of recovered metal that was found to be mostly aluminum with some iron, a finding that they declare to be anomalous since aluminum alloys usually contain copper. But two minutes on Google reveal that aluminum iron alloy is widely available from any number of suppliers, in various purities up to 99.999%, and with virtually any ratio of aluminum to iron that you want. MUFON also declined to give any information about who performed the alleged analysis.

They make much of the sample's appearing to have been exposed to heat, which they attribute to the alleged explosion; but who knows what engine block or other piece of hot machinery this piece may have been a part of. MUFON's whole report reads (to me) as if they're starting from the presumption that a spaceship exploded over Judge Proctor's property, and they're looking for things that might be consistent with their "artist's impression" of such an event.

So if we want to know what really happened in Aurora in 1897, we're going to have to do it without the obvious testable evidence. We're going to have to dig not into the ground, but into the newspapers.

It turns out that for four days prior to the Aurora crash, Texas newspapers had been abuzz with reports of balloons and airships. By 1897, when this happened, hot air balloons had been flying for more than a century, and their appearance over Texas may well have been a curiosity but hardly unknown. But Texas newspapers were in the habit of outdoing each other. The *Texas Almanac* gives this account of 38 newspaper reports covering 23 counties:

> *Newspapers of the day reported the sightings straight-faced, although one can read more than a little tongue-in-cheek writing into some of the dispatches from community correspondents... In Farmersville, an eyewitness saw three men in the cabin and heard them singing "Nearer My God to Thee."*

> *The trio reportedly also was passing out temperance tracts... Texans always have to one-up each other, and the "airship" craze provided a perfect setting.*

And so it was only after nearly a week of this colorful storytelling that the *Dallas Morning News* decided to run a full page featuring the best of all these stories. Page 5 of the April 19, 1897 edition contained the editors' sixteen favorite yarns, each a silly story from a different town. In one, an "aerial monster" landed in a field, piloted by men from New York. In another, the crew consisted of lost Jews from the ten tribes of Israel who told a judge they'd come from the North Pole. Another of the stories was the following, given here in its entirety:

> *About 6 o'clock this morning the early risers of Aurora were astonished at the sudden appearance of the airship which has been sailing through the country.*
>
> *It was traveling due north, and much closer to the ground than ever before. Evidently some of the machinery was out of order, for it was making a speed of only ten or twelve miles an hour and gradually settling toward the earth. It sailed directly over the public square, and when it reached the north part of the town collided with the tower of Judge Proctor's windmill and went to pieces with a terrific explosion, scattering debris over several acres of ground, wrecking the windmill and water tank and destroying the judge's flower garden.*
>
> *The pilot of the ship is supposed to have been the only one on board, and while his remains are badly disfigured, enough of the original has been picked up to show that he was not an inhabitant of this world.*
>
> *Mr. T. J. Weems, the United States signal service officer at this place and an authority on astronomy, gives it as his opinion that he was an inhabitant of the planet Mars.*
>
> *Papers found on his person — evidently the record of his travels — are written in some unknown hieroglyphics, and can not be deciphered.*
>
> *The ship was too badly wrecked to form any conclusion as to its construction or motive power. It was built of an unknown*

metal, resembling somewhat a mixture of aluminum and silver, and it must have weighed several tons.

The town is full of people to-day who are viewing the wreck and gathering specimens of the strange metal from the debris. The pilot's funeral will take place at noon to-morrow.

Why the UFOlogists decided that this one story out of all the dozens was a literal true account, while the others were obvious jokes, is a secret known only to themselves. Author Bill Porterfield spent some time in Aurora in 1973 while the UFOlogists were doing their investigation. Porterfield checked up on the people mentioned in the news story, and he detailed what he learned in one chapter of his 1978 book *A Loose Herd of Texans*.

The 1897 article was written by S. E. Haydon, a cotton buyer in Aurora, left over from the town's boom times prior to 1890. Judge J. S. Proctor, upon whose property the crash took place, was a local justice of the peace and a friend of Haydon's. Both men often submitted satirical essays and poems to local papers. Proctor even wrote his own version of Haydon's alien tale and published it in his own paper called the *Aurora News*, and when town constable J. D. Reynolds read it, he "roared with laughter" and said "The judge has really outdone himself this time."

Porterfield even tracked down T. J. Weems, described in Haydon's article as a "United States signal service officer... and an authority on astronomy." This was, evidently, a joke on a friend; as T. Jeffrey Weems was Aurora's blacksmith and farrier, and knew no more about astronomy than his average equine customer.

Along with Haydon and Proctor, Weems had been about the only businessman left in the town. The 1890s had been cruel to Aurora. Two epidemics, reported as either yellow fever or spotted fever, had killed many, including Haydon's wife and two of his four sons. The Burlington Northern Railroad canceled plans to extend the Rock Island line to Aurora. A boll weevil infestation decimated Aurora's principal industry, cotton, thus ravaging Haydon's business. Finally a fire destroyed much of Aurora's small downtown. By 1897 nearly everyone still alive had left, leaving only a few hardy souls. Porterfield concluded that what sustained Haydon and Proctor was their sense of humor.

All the evidence points to the story being nothing more than Haydon and Proctor's contribution to a running joke that swept Texas one spring, a long time ago. The other evidence we'd expect to exist if the story were true — the "several tons" of wreckage, the papers written in hieroglyphics, reports of the funeral taking place, records of T. J. Weems actually being an Army officer — conspicuously do not exist. What does exist is a small town, unimportant to all but those who live there, that's exactly as we'd expect it to be if no aliens ever crashed into Proctor's apocryphal windmill and cost him his flower garden.

That is indeed a small story, to be sure. It's a glimpse into a few sturdy men with dashed dreams. It's a boom that went bust and hopes that never came to be. The one element common to both versions of Aurora's history is that each ended in the town's cemetery. Porterfield was never able to track down what finally became of Haydon or Proctor, but in one way or another, they're in the cemetery too; either in body, or in the spirit of the joke they concocted to make some hard times just a little easier.

REFERENCES & FURTHER READING

Goodwyn, M. *Chasing Graveyard Ghosts: Investigations of Haunted and Hallowed Ground.* Woodbury: Llewellyn Publications, 2011. 93.

Hays, R. *Military Aviation Activities in Texas, World Wars I and II.* Austin: University of Texas, 1963.

Jack, A. *Loch Ness Monsters and Raining Frogs: The World's Most Puzzling Mysteries Solved.* New York: Random House, 2009. 3-11.

MUFON. "Aurora Texas Crash." *MUFON Case File.* Mutual UFO Network, 1 Jan. 1973. Web. 13 Jan. 2011. <http://www.mufon.com/famous cases/Aurora+Texas+Crash+Part+1 +MUFON+Case+File.pdf>

Peralta, E. "Can a Space Alien Rest in Peace?" *Houston Chronicle.* Hearst Newspapers, 1 Mar. 2007. Web. 10 Jan. 2011. <http://www.chron.com/disp/story.mpl/chronicle/4587362.html>

Porterfield, B. *A Loose Herd of Texans.* College Station: Texas A&M University Press, 1978. 174-184.

Texas Almanac. "When Airships Invaded Texas!" *Early Balloons in Texas.* Lloyd Cates, 1 Jan. 2006. Web. 11 Jan. 2011. <http://lloydcates.squarespace.com/early-balloons-in-texas/>

50. Nuclear War and Nuclear Winter

Will atmospheric smoke from a nuclear war result in devastating global cooling?

Put on your jackets; in this chapter, we're going to prepare for the nuclear winter, the theorized period of catastrophic global cooling following a nuclear war. Some say that, if war happens, a nuclear winter is a certainty that will devastate agriculture and kill billions; some say that it's greatly overstated or even outright made-up by anti-nuke activists. Does the prospect of a nuclear winter truly constitute one more reason why nuclear arsenals should be dismantled?

Our purpose today is not to explore the myriad other implications of nuclear war, or to otherwise prove that it's a bad thing. That's pretty clear. The obvious effects of a nuclear blast, the initial explosion itself and the radioactive fallout, are well established and not in dispute. As destructive as those are, their long-term environmental effects are negligible. So what, then, causes the nuclear winter?

The concept of a nuclear winter entered the mainstream in 1983, when the "TTAPS" team, named for its authors Richard Turco, Owen Toon, Thomas Ackerman, James Pollack, and Carl Sagan, reviewed existing work and ran computational climate simulations to see what would happen when huge amounts of smoke were added to the atmosphere. The source of this smoke is not the nuclear explosion itself, but the all the building fires and wildfires that would follow each one. Take 150 nukes striking major population centers worldwide, and that's a lot of fires, with a smoke output greatly exceeding anything in human history. Members of the TTAPS team have also published many followup papers, revising and improving their estimates, but generally with similar results.

Almost all of the simulations run by scientists replicating these results agree, at least in broad strokes, when the input variables are the same. It should be stressed that there are a lot of these variables. The principal weakness that these studies all share hinges on one of these variables, and it's a very important one: Exactly how much smoke will the fire following a nuclear bomb produce? In 1986, Joyce Penner from the Lawrence Livermore National Laboratory published an article in the journal *Nature* in which she pointed out that this specific variable is responsible for determining whether the effects will be minor or massive. She also found that the published estimates of this varied widely.

Penner's paper was not the only one critical of the TTAPS predictions. However, this fact is often misinterpreted. The differences that have been found among the various simulations are of degree. Laypeople who hear of the criticism often think that the idea of a nuclear winter has been "debunked" or that it's some kind of discredited myth. This is not the case. Perhaps the most often cited and most critical paper was "Nuclear Winter Reappraised" by Starley Thompson and Stephen Schneider, published in 1986 in the journal *Foreign Affairs*. They did not dismiss the idea at all; rather they re-characterized it as a nuclear autumn.

So here we get into the meat of the question. We know with pretty good certainty how a given amount of smoke in the atmosphere, distributed a certain way, will affect the climate and for how long; but what we can only guess at is how much smoke is produced when a city burns after a nuke. Our guesses are educated, but they're all over the map. Cities also vary wildly in just about every relevant aspect. Let's look at what we know from history.

An obvious question to ask is whether these effects have been seen with any of the nuclear tests that many nations have conducted. Some 2,000 nuclear bombs have been detonated, somewhat less than half of which were in the atmosphere and are comparable to what would be used in a war. In none of them were any harmful smoke-induced environmental effects produced. However the reason for this is quite simple. Nuclear tests are not performed in cities filled with tens of thousands of combustible buildings; they happen way out in the desert or over the ocean, and no subsequent fires are created.

But what about the two cases when atomic weapons were used on real-life cities, Hiroshima and Nagasaki? Discussion of the subsequent fires in both cities are hard to come by, as they were not really what people were focusing on. Hiroshima developed a firestorm — where it builds into a single large fire with a central heat core that draws in oxygen with a powerful wind from all around — that peaked two to three hours after the explosion. Six hours after the explosion, nearly everything combustible within a one-and-a-half kilometer radius had been consumed, and the fire was almost completely out, leaving over 8 square kilometers destroyed. Descriptions of residual and secondary fires outside the radius of the firestorm are rare and hard to find, but it seems likely that several hundred or thousand small fires continued for the better part of 24 hours. Photographs taken of Hiroshima over the next few days do not show any significant evidence of vast amounts of smoke.

Nagasaki was hit with a larger bomb, but its geography spared it a firestorm. Whereas Hiroshima is centered in a large flat plain, Nagasaki is irregularly shaped among hills and valleys, and cleft by a large harbor. Secondary fires were widespread, and Nagasaki firefighters had to cope with a damaged water system. It took several days to get the many small structure fires controlled or burned out. But Nagasaki's geography meant that there were far fewer fires than in Hiroshima. Again, the post-nuke photographs don't show vast atmospheric plumes of smoke.

When the Iraqi army set 700 of Kuwait's oil wells on fire when they retreated in 1991, the wells burned for eight months, lofting about a million tons of smoke into the atmosphere. The TTAPS team predicted global climate change effects, that fortunately failed to materialize. Carl Sagan discussed this error in his book *The Demon-Haunted World*, and later research discovered the reason. The smaller individual smoke plumes, spread over a wide area, did not generate sufficient uplift to get the smoke into the upper atmosphere, even though theoretically enough smoke was produced. Temperatures did drop over the Persian Gulf, but the effect remained localized.

Other cataclysmic events have proven that the nuclear winter scenario is not at all far-fetched. The eruption of Mt. Pinatubo in the Philippines, also in 1991, threw some 17 million tons of particulates into the upper atmosphere that caused global temperatures to drop by about a degree for several months. Sunlight dropped by 10%. This temperature drop did not, however, have any long-term effect on agriculture.

Pinatubo was only a blip compared to the K-T extinction event of some 65 million years ago, when a theorized asteroid hit us with one hundred million megatons of destructive force, lighting virtually the entire world on fire. The evidence of this is called the K-T boundary, a layer of clay found all around the world. Sunlight was reduced by 10-20% for ten years, which caused a massive cascading extinction of species from plants to herbivores to carnivores.

But we shouldn't expect anything like this to happen from a nuclear war. Times continue to change, including the nature of warfare. Nations no longer stockpile the megaton class weapons popular in the 1950s and 1960s; typical yields now are a fraction of a megaton. The United States' conventional capability is now so good that it can effectively destroy an entire nation's ability to wage large-scale war overnight, using only conventional weapons. But that doesn't mean the nuclear forces are no longer needed. Should a superpower strike first against the United States with nuclear weapons, the response would more than likely be nuclear, bringing Mutually Assured Destruction into play. But what about a small nation striking first? What about nukes in the trunks of cars parked in major cities? In the modern era, it's much less clear that any superpower would necessarily have anyone to shoot back at.

Increasingly, non-superpower nations are building nuclear stockpiles. India and Pakistan might get into it with one another. Israel's foes might surprise it with nuclear weapons. Who knows what North Korea and Iran might do. Smaller regional nuclear wars remain a very real possibility. According to the worst-case estimates in the TTAPS papers, about one million tons of smoke would be expected from the fires resulting from each nuclear strike. And these smaller regional nuclear combats are expected to use about 50 nuclear weapons (compare this to 150 nuclear weapons for a broader global nuclear war). Thus, today's most likely nuclear scenario would be expected to produce climate effects similar to three Pinatubo events, according to the worst estimates, and still many orders of magnitude less than the K-T extinction.

And so, while the nuclear winter scenario is a good prediction of the effects of a worst-case scenario, when all the variables are at their least favorable, the strongest probabilities favor a much less catastrophic nuclear autumn; and even those effects depend strongly on variables like whether the war happens during the growing season. A bomb in Los Angeles might result in history's worst firestorm, while a bomb in the mountains of Pakistan might create no fires at all. The simple fact is that there are too many unpredictable variables to know what kind of climate effects the smoke following nuclear fires will produce, until it actually happens. Obviously we're all very mindful of the many terrible implications of nuclear combat, and if it ever happens, the prospect of a nuclear autumn will likely be among the least of our concerns. The physicist Freeman Dyson perhaps described it best when he said "(TTAPS is) an absolutely atrocious piece of science, but I quite despair of setting the public record straight... Who wants to be accused of being in favor of nuclear war?"

References & Further Reading

American Geophysical Union. "Spacecraft Provides First Direct Evidence: Smoke In The Atmosphere Inhibits Rainfall." *Science News.* ScienceDaily, LLC, 6 Oct. 1999. Web. 7 Feb. 2011. <http://www.sciencedaily.com/releases/1999/10/991005125625.htm>

Child, J. *Nuclear War: The Moral Dimension.* New Brunswick: Transaction Books, 1986. 69.

Penner, J. "Uncertainties in the smoke source term for 'nuclear winter' studies." *Nature.* 20 Nov. 1986, Volume 324: 222-226.

Sagan, C. *The Demon-Haunted World: Science as a Candle in the Dark.* New York: Random House, 1995. 257.

Thompson, S., Schneider, S. "Nuclear Winter Reappraised." *Foreign Affairs.* 1 Jun. 1986, Volume 62: 981-1005.

Turco, R., Toon, O., Ackerman, T., Pollack, J., Sagan, C. "Nuclear Winter: Global Consequences of Multiple Nuclear Explosions." *Science.* 1 Jan. 1983, Volume 222, Issue 4630: 1283-1292.

Index

2001: A Space Odyssey, 118
2012, 100, 103
9/11 attacks, 107, 110, 182, 210
Ackerman, Thomas, 344
Acquired Immune Deficiency Syndrome (AIDS), 271
acupuncture, 71, 96
Africa, 46, 54, 58, 81, 178, 226, 229, 283, 284, 300
Air France Flight 447, 293, 295, 296, 297, 298
Airborne (supplement), 267
Airbus, 296
Aldrin, Buzz, 219
aliens, 2, 4, 34, 36, 37, 39, 46, 47, 48, 51, 100, 101, 104, 134, 136, 138, 174, 213, 217, 218, 219, 220, 221, 222, 288, 337, 341, 342
Alistair Munro, 157
Altiplano, 134, 137
aluminum, 26, 29, 339, 341
Alyn, Kirk, 321
American Psychological Association, 303
amero (currency), 111, 183, 184, 185
amino acids, 26, 203, 204, 206, 253
Amish, 151
Amway, 9
Andes Mountains, 134, 286, 287, 288
Anspaugh, Lynn, 323

antifreeze, 28, 252
Antikythera Mechanism, 46, 47, 48, 49, 50
Anza, Juan Bautista de, 178
Anza-Borrego Desert, 176
apocalypse, 100, 103, 105
Apollo program, 218, 219, 295
Appalachian Mountains, 258, 259, 264
aqua tofana, 199, 200
Archimedes, 48, 49
Argentina, 149, 286
Argentinian Civil Aviation Accident Investigation Board (JIACC), 287
Armstrong, Neil, 219
Army Research Institute, 243
aspartame, 30
Astor Place Riot, 246, 247
astrology, 48, 109, 238, 241
Atlantis (mythical city), 47, 231, 232, 236
Attaran, Amir, 283
Attention Deficit Hyperactivity Disorder (ADHD), 328
aura, 142
Aurora, TX, 337, 339, 340, 341, 342
Auschwitz, 102, 148, 149, 150
autism, 325, 328
autoimmune diseases, 266, 268, 327

autolysis, 254
Avro Lancaster (aircraft), 287
Avro Lancastrian (aircraft), 286, 287
Avrocar (aircraft), 170
Aztecs, 126, 128, 129, 132
BadAstronomy.com, 331
Bai Yu, 35, 36, 38, 39
Baigong Pipes, 2, 34, 35, 36, 37, 38, 39
bald eagles, 282, 283
ball lightning, 87, 88, 89, 90, 91, 92
barefoot, 2, 53, 54, 55, 56, 57, 58
barley, 325
Barrett, Stephen, 143
bathymetry, 309, 310
Beethoven, Ludwig van, 197, 198
Begin, Menachem, 80
Bell Curve, The (book), 302, 303
Bell Island Boom, 73, 74, 75, 77, 78
Bellevue Hospital Center, NY, 301
Bélmez, Spain, 93, 94, 96, 97, 98
Ben-Hur (movie), 313, 314
Berg, Howard, 274, 277
Bering Strait, 178
Bernifal Cave, 164
Big Bang, 15
bigfoot, 4
Bikila, Abebe, 54
Binet, Alfred, 301, 304

Binet-Simon test, 301
Blann, Tommy, 337
Bloch, Claude, 192
Boelza, Igor, 197
boll weevil, 342
Bolsheviks, 186
Borman, Frank, 221
Botts, Myrtle, 176, 177, 179
Braille, 104
Brazil, 147, 149, 150, 152, 296
Briggs Myers, Isabel, 238
Briggs, Katherine, 238
British Airways Flight CS-59, 286, 287, 288, 289, 290
British Broadcasting Corporation (BBC), 232
bronze, 46, 47, 48, 49, 50
Brown Mountain Lights, 258, 259, 260, 261, 262, 264
Brynner, Yul, 80, 84
Buckland, William, 255
Budd, Zola, 54, 58
Buddhism, 40, 41, 42, 162
Buenos Aires, Argentina, 149
bullshido, 67, 68, 69, 70, 71
Bureau of Investigations and Analysis for the Safety of Civil Aviation (France), 296, 297
Burlington Northern Railroad, 341
Cable News Network (CNN), 186
Cain, Dean, 321
Callahan, Philip, 131
calories, 252, 253

Cámara, María Gómez, 93, 94, 95, 97
Camarasa, Jorge, 147, 150
Cameron, Harry, 68
cancer, 4, 13, 14, 64, 266, 271, 284, 321, 322, 323
Cândido Godói, Brazil, 147, 150, 151, 152
carbohydrates, 325, 329
cargo cults, 120, 123, 124
Carrizo Badlands, 177, 179
Carson, Rachel, 280, 281, 282, 284
Carter, Jimmy, 185, 281
Carver, Ronald, 274
Case, Bill, 337
Catawba nation, 263
Catholic church, 94, 128
celiac disease, 327, 328
cellular phones, 13
Centers for Disease Control, 26, 30, 158
Charlotte, NC, 258, 262
Checklin, Donald, 288
chemtrails, 156
Cherokee nation, 261, 263
chimpanzees, 226, 229
China, 35, 36, 165, 178, 229, 231, 279
Christian, Robert, 114, 116
Cipac de Aquino, Marcos, 129
Clarke, Arthur C., 118, 120
Clausen, Henry, 190
Codex Escalada, 130
coffee, 14, 27
collagen, 17

Collins, Michael (astronaut), 219
Collyer, Bud, 321
Colonias Unidas, Paraguay, 150
Colorado Desert, 176, 179
Colorado River, 176, 177, 178
Commonwealth Aircraft Corporation, 169
commotio cordis, 70
computed tomography, 50, 308
Concorde (aircraft), 76
concussion, 70
Condon, Edward, 220, 221
Connery, Sean, 314
Conqueror, The (movie), 321, 322, 323
conspiracy theories, 45, 51, 53, 73, 75, 99, 100, 103, 104, 107, 109, 110, 111, 112, 113, 114, 116, 117, 119, 156, 172, 174, 182, 183, 185, 186, 187, 188, 189, 190, 192, 193, 198, 211, 214, 229, 233, 295, 316
Consumer Reports, 8
Convair, 75
Cook, Reginald, 288
corona discharge, 90
Cortés, Hernán, 128, 129
cotton, 341, 342
Cro Magnon, 164
Crow, The (movie), 316
cryptozoology, 162, 163, 165, 229
cuvieronius, 138

Dallas Morning News, 340
Dandenong Journal, 171, 172
Danson, Ted, 319
Dapendeng culture, 235
Dark Side of the Moon (album), 318
Darwin, Charles, 300
Dassault Mirage III, 169
Daumer, Georg Friedrich, 199
Dawson, Charles, 254
DDT (dichlorodiphenyltrichloroethane), 279, 280, 281, 282, 283, 284
de Brahm, John William Gerard, 261, 263, 264
de Gante, Pedro, 129
De Havilland Australia, 169
Dean, James, 317, 318
Death Valley, 138
Deepwater Horizon, 182
dehydration, 281
Denver International Airport, 99, 101, 102, 104, 105
diabetes, 17, 284
Diaz, Melchior, 178
Diego, Juan, 126, 127, 128, 130, 131, 132
dielectric effect, 18
Dillman, George, 68
dim mak, 67, 69
dinosaurs, 161, 163, 164, 165, 166, 255
Dionysius of Alexandria, 48
disaccharides, 204
Discovery Channel, 228, 232

Disney, Walt, 319
DNA, 16, 148, 150, 178
Dobbs, Lou, 186
dolphins, 306, 307, 311
downwinders, 322
Dunne, Dominique, 315
Dyson, Freeman, 348
earthquakes, 176, 232, 235
Edward, Mark, 21, 23
Egypt, 80, 81, 82, 83, 84, 85, 165, 234
Eichmann, Adolf, 149
Elberton, Georgia, 113, 114, 115, 116, 117, 118
Electromagnetic Biology and Medicine (journal), 17
electromagnetic pulse, 74, 78, 100
electromagnetic theory, 88, 142
Elephantine Island, 82, 83
Elizabeth II, Queen, 100, 105
Ellison, Arthur, 20
Environmental Protection Agency (EPA), 281, 284
Environmental Protection Information Center, 307, 308
Escalada, Javier, 130
Esson, Dylan James, 323
Estabany, Oskar, 141
estivation, 252, 253, 256
eugenics, 114, 300, 302, 304
euro (currency), 183
European Space Agency (ESA), 333

European Union (EU), 183, 186
evolution, 53, 54, 55, 56, 57, 111, 124, 202
Exodus, 80, 82, 84, 85
Extremadura, Spain, 128, 129
farming, 82, 135, 137, 213
fatty acid, 204, 253
Federal Emergency Management Agency (FEMA), 182
Federal Trade Commission (FTC), 8, 11, 267, 274
Fendley, Joe, 114, 115, 116, 117, 118
fish, 45, 280
Flynn effect, 304
Flynn, James, 304
Fontana, David, 20
Food and Drug Administration (FDA), 267
Forer Effect, 241
Forer, Bertram, 241
formaldehyde, 26, 27, 31
Forrest, Edwin, 247
founder effect, 151
Fox, Vicente, 184, 185
Foy, Robin and Sandra, 23
Franciscans (Catholic order), 129, 130, 131
Freemasonry, 100, 104, 117, 198
Fresco, Jacques, 108
Freyman, Robert, 75, 76, 77
Friendship 7, 217
frogs, 45, 251, 252, 253, 254, 255, 256

fructose, 204
Frum, John, 121, 122, 123
Galilei, Galileo, 333
Galileo (spacecraft), 333
Galton, Sir Francis, 300
gamma rays, 15
Gauld, Alan, 23
Gemini program, 218, 220, 221, 222
Georgia Guidestones, 113, 114, 115, 117, 118, 119
Gezer Calendar, 83
ghosts, 4, 212, 213, 245, 258, 261, 319, 336
Giza Necropolis, 82
Glenn, John, 217
Global Position System (GPS), 335
glucose, 204, 252
gluten, 325, 326, 327, 328, 329
gluten free diets, 325, 326, 327, 328, 329
gluten sensitivity, 327, 328, 329
Google Earth, 102, 335
Grange Reserve, 169, 171
gravity anomaly, 331, 332, 336
Gravity Hill, PA, 334, 335
gray aliens, 4
Great Pyramid of Cheops, 82, 84, 234
Great Smoky Mountains, 263
Green party, 14
Greenpeace, 283

Greenwood, Andrew, 170, 173
Gregory, C. E., 263
Grubel, Herb, 111, 184
Guinness Book of World Records, 274
Handel, Paul, 89, 90
Harmer, Dennis, 287, 288
Havas, Magda, 16, 17
Haydon, S. E., 341
Hayward, Susan, 321
Hebrews, 80, 81, 83
hepatitis, 103
Herbalife, 9
Here Be Dragons (film), 107
Herodotus, 83
Herrnstein, Richard, 302, 303
Heston, Charlton, 80, 313
Hewes, Hayden, 337
hibernation, 252, 253
Hill, Betty and Barney, 4
Hilton-Cook, Norman, 288
Hipparchos, 49
Hiroshima, Japan, 346
Hirsch index, 215
History Channel, 232, 337
Hitler, Adolf, 147, 213
Hoagland, Richard, 112
Hollywood, 80, 219, 313, 317, 319, 323
Hollywoodland (movie), 320
Holocaust, 81, 149
Holy Bible, 81, 103
homeopathy, 96
Horus (Egyptian god), 109
Hubbard, L. Ron, 116

Humanzee (documentary film), 228, 229
Hunsrück, Germany, 152
hydrochloric acid, 29
Iceland, 178
Illuminati, 99, 100, 101, 102, 103, 104, 105
immune system, 27, 30, 32, 103, 254, 266, 267, 268, 269, 270, 271, 327
impact factor, 215
Inca, 134
India, 48, 136, 279, 348
International Agency for Research on Cancer, 13
International Space Station (ISS), 295
intervention, 63, 64, 145, 271
Inuit, 178
iPhones, 48, 264
IQ testing, 299, 300, 302, 304
Israel, 81, 83, 340, 348
Israelites, 81, 83
Ivanov, Il'ya Ivanovich, 224, 225, 226, 227, 228, 229
James, King, 245, 248
Japan, 69, 188, 189, 190, 192, 193, 231, 232, 236, 293, 309
Javan rhinoceros, 164
Jesse, Edward, 256
Jesuit (Catholic order), 130
Jet Propulsion Laboratory, 131
Jews, 80, 81, 83, 84, 85
JN-25 (Japanese cipher), 190
Johns Hopkins, 155

Johnson Space Center, 220
Johnson, Lyndon, 124
Jones, Alex, 112
Jorgensen, Chuck, 276
Joseph II, Emperor, 196, 200
Jung, Carl, 238, 240, 241, 243
Kamchatka, Russia, 231
Kanoksilp, Manos, 43
Katrina (hurricane), 182
Keen, Montague, 20
Kennedy, John Fitzgerald, 274, 277
Ketten, Darlene, 308
Kimmel, Husband, 189, 190, 191, 192
Kimura, Masaaki, 232, 233, 234, 235
Kondos, Dimitrios, 46
Krieger, Dolores, 141, 142
Krishna (Indian god), 109
K-T extinction, 347, 348
Kunz, Dora, 140
kyusho jitsu, 67
Lai, Henry, 17
Lake Cahuilla, 177, 178, 179
Laos, 40, 41, 42, 45
Las Vegas, 11
latex, 29, 314
Lawrence Livermore National Laboratory, 345
Lee, Brandon, 316
Lee, Bruce, 67, 316
Leitao, Mary, 155, 157
leukemia, 322
leukocytes, 268, 269
Levi, Primo, 80

lightning superbolts, 76, 77, 78
Liszt, Franz, 198
logical fallacies, 209, 215
Loose Change (film), 107, 112
Los Alamos National Laboratory, 75
Los Angeles Times, 321
Lovell, James, 221, 222
Lowke, John, 90
Maachi MB-326, 169
MacDonald, Gordon, 75
Macready, William, 246
Magic Kingdom (theme park), 319, 320
malaria, 279, 283, 284
Manehivi, 122, 123
Marfa Lights, 258
martial arts, 67, 68, 69, 70, 71, 316
Martin, H. C., 262
Martin, Wyatt, 114
Mayans, 100, 103
McCarthy, Jenny, 31, 212
McDivitt, James, 220, 221
Mekong River, 40, 41, 42, 43, 45
Melbourne, Australia, 168, 169, 170
Mengele, Josef, 147, 148, 149, 150, 151, 152
mercury (element), 28, 200, 281
Mercury program, 28, 218
Merlin (engine), 287, 290
Merola, Peter Joseph, 108, 109, 110, 111

metabolism, 28, 204, 252, 253, 254
Mexicali, 177
Miklouho-Maclay, Nicholas, 124
Min Min light, 258
Miocene epoch, 180, 234
mirage, superior, 258, 262
Mitchell, Edgar, 218
Mithra (Persian god), 109
mitochondrial Eve, 151
Mojave Desert, 331
Mokele Mbembe, 161, 162, 166
Mona Vie, 9
monosaccharides, 204
Monster Quest (TV show), 228
Montúfar, Alonso de, 129, 130
Moorabbin Airport, 169
Moorehead, Agnes, 321
Morgellons Disease, 154, 155, 156, 157, 158, 159
Morris, Desmond, 229
Morse code, 286, 290, 291
Mozambique, 283
Mozart, Constanze, 199, 200
Mozart, Wolfgang Amadeus, 195, 196, 198, 199, 200
Mt. Pinatubo (volcano), 347, 348
multilevel marketing, 7, 8, 9, 10, 11, 60
multiple sclerosis, 266, 329
Murray, Charles, 302
Mustang (aircraft), 287

mutually assured destruction, 348
Myers-Briggs Type Indicator (MBTI), 238, 239, 240, 241, 242, 243
Mystery Spot, The (CA), 332
mystery spots, 331, 332, 333, 334, 336
Mythbusters (TV show), 18
naga, 40, 41, 42, 43, 44, 45, 162
Nagasaki, Japan, 346
Nahuatl, 127
National Aeronautic and Space Administration (NASA), 217, 218, 219, 220, 221, 276
National Cancer Institute, 322
National Enquirer, 228
National Oceanic and Atmospheric Administration (NOAA), 297, 311
National Science Foundation, 51
National Transportation Safety Board (NTSB), 102, 296
Nature (journal), 345
Navajo, 104
Navy, Tom, 121
Nazism, 99, 100, 147, 148, 149, 150, 213, 214, 300
Nevada Test Site, 321
New England Journal of Medicine, 322
New Hebrides, 121, 122, 123

New World Order, 99, 100, 105, 114, 117, 119
New York Post (newspaper), 321
Newfoundland, 73, 74, 78
Newsweek, 9
Nican Mopohua, 127, 130
Nike (sportswear), 55
Nile River, 80, 82
Nimitz, Chester, 190
Nissen, Georg Nikolaus von, 199
North American Free Trade Agreement, 184
North American Union, 182, 183, 185, 186, 187
North Korea, 279, 348
Nubia, 83
nuclear autumn, 345, 348
nuclear winter, 344, 345, 347, 348
Nuremberg Trials, 300
O'Rourke, Heather, 315
obesity, 270, 325
Ocean Realm (magazine), 44
Ogopogo, 161
Oliver (chimpanzee), 229
omega-3 fatty acid, 204
omega-6 fatty acid, 204
Oregon Vortex, 332
organic food, 28, 202, 205, 328
Ornate, don Juan de, 178
oxidation, 253, 254
Pacific Ocean, 176, 178
Pakistan, 348
Palace of Darius, 136

paleohydrology, 137, 177
papyrus, 83
parasites, 154
Parsons, Alan, 319
Parthenon, 136
Passover, 83
Pastor, Robert, 111, 185
pathogens, 27, 32, 269, 270
peanut allergies, 329
Pearl Harbor, 188, 189, 190, 191, 192, 193
Peek, Kim, 273, 274, 276
pelicans, 282
Pendleton, Robert, 321
Penn & Teller, 22
Penner, Joyce, 345
People (magazine), 321, 322
Peralta, Alberto, 131
Persepolis, 136
Persian Gulf, 347
pesticides, 279, 283, 284
pH, 29
Phayanak, 40, 41
phenylalanine, 203
phenylketonuria, 203
Phon Phisai (Thailand), 40, 41, 42
Piltdown Man, 254
Pink Floyd, 318
Pisgah National Forest, 258
placebo effect, 60, 159, 328
Plait, Phil, 331
plasma, 77, 89, 90, 91
Pliocene epoch, 180
plutonium, 14
polio, 32

Pollack, James, 344
Poltergeist (movie), 315
polysaccharides, 204
Ponzo Illusion, 334
porpoises, 307
Porsche 550 Spyder, 317, 318
Porterfield, Bill, 341
Proctor, J. S., 338, 339, 340, 341
Psammetichus I, 83
PubMed, 242
Pumapunku, 134, 135, 136, 137, 138
Purple (Japanese cipher), 190
Pushkin, Alexander, 196
pyramids, 82, 231
Pyrex, 89
Qaidam Basin (Tibet), 34, 35, 37, 38
qi, 69, 71
Radiation Exposure Compensation Act, 322
radiation, ionizing, 15
radiation, microwave, 89
radiation, radio frequency (RF), 13, 17, 77
radiation, ultra high frequency (UHF), 89
radiofrequency, 13, 14, 15, 18
radiofrequency identification (RFID), 77
Ramesses I, 84
Ramesses II, 80, 84
rauding, 276, 277
Rayner, Keith, 275
Reeve, Christopher, 320
Reeves, George, 320

Reeves, Richard, 277
Rennie, John, 1, 3
reptoids, 101
Reynolds, J. D., 341
rickshaws, 54, 58
Rimsky-Korsakov, Nikolai, 196
Robins, Oliver, 315
Roosevelt, Franklin D., 190, 193
Ropen, 161, 162, 166
Rosa, Emily, 143, 145
Rosenkreuz, Christian, 116
Rosicrucianism, 116
Rossini, Gioachino, 198
Royal Australian Air Force, 171, 172
Russian Revolution, 186, 225
rye, 325
Ryukerin, Kiai Master, 68
Ryukyu Islands, 235, 236
Sagan, Carl, 111, 344, 347
Sahagún, Bernardino de, 131
Saint Ignace Mystery Spot (MI), 332
Salieri, Antonio, 195, 196, 197, 198, 200
Salton Sea, 177
Salton Sink, 177, 179
Samblebe, Frank, 172
Sánchez, Eduardo Chavez, 128
Sauthier, Paolo, 152
Schenider, Stephen, 345
Schoch, Robert, 234
Schubert, Franz, 198

Scientific American (magazine), 3, 88, 89, 90, 143
Scole Experiment, 20, 21, 22, 23, 24
Scripps Institute, 44
Sea of Cortez, 176
séances, 20, 21, 22, 23, 24, 25
semites, 81
Serino, Danielle, 68
Shakespeare, William, 245, 248, 249
sheeple, 105
Shellmound phase, 235
Shroud of Turin, 126
Silent Spring (book), 280, 281, 283
Simon, Théodore, 301
skepticism, 1, 2, 5, 61, 187
Sky Scorcher, 75
Skylab, 217
smart meters, 14, 18
Society for Psychical Research, 20, 22, 24, 214
solar radiation, 297
sonar, 306, 307, 308, 309, 310, 311
South Atlantic Anomaly, 293, 294, 295, 296, 298
Soviet Union, 75, 170, 186, 218, 221, 224, 225, 226, 227, 228
speed reading, 273, 274, 275, 276, 277
Sphinx (Egypt), 82, 234
Spielberg, Steven, 317
Spitfire (aircraft), 287
St. Elmo's Fire, 90

Stalin, Joseph, 224, 227, 228, 229
Stanford University, 301
Stanford-Binet test, 301, 302
Star, The (tabloid), 321
Stegosaurus, 162, 163, 164, 165
STENDEC, 286, 287, 288, 289, 290, 291
Stinnett, Robert, 190
Straight Dope, The, 277
stratosphere, 76
Strauss, Levi, 7
stress, 154, 156, 158, 159, 281, 308
subvocalization, 275, 276
sugar, 17, 204, 206
Superman (various iterations), 320, 321
supersonic flight, 76
Süssmayr, Franz Xavier, 198
swamp gas, 42, 45, 261
Syracuse, 49
Ta Prohm, Cambodia, 162, 163, 164, 165
Table Rock, NC, 260
Taiwan, 231, 235
Tanguma, Leo, 102, 103
Tanna Law, 122
target drogue, 171, 174
Tarzan (orangutan), 226
Taylor, Barbara, 311
Tepeyac (hill), 126, 129, 130
Ternan, Lewis, 301
Tesla, Nikola, 75, 91
Texas Almanac (newspaper), 339

The New Yorker (magazine), 280
therapeutic touch, 140, 141, 142, 143, 144, 145
thimerosal, 26, 28
Thompson, Starley, 345
thoughtography, 93
Three Men and a Baby (movie), 319
Tibet, 34
tilma, 127, 130, 131, 132
Titicaca (lake), 134, 136, 137, 138
Tiwanaku, 134, 135, 136, 137, 138, 139
tobacco, 14
Tonantzin (Aztec goddess), 129, 132
Toon, Owen, 344
toxodon, 138
Trent University, 16
Trudeau, Kevin, 274
TTAPS, 344, 345, 347, 348
Tupungato Mountain, 286, 288
Turco, Richard, 344
Tutankhamen, 248
Twain, Mark, 1
Tyack, Peter, 311
Tyler, Kermit, 189, 193
typhus, 279
Udayagiri Caves, 136
UFO Files (TV show), 337
UFOlogy, 337, 338, 341
ultraviolet radiation, 14, 15

unidentified flying objects, 4, 42, 168, 169, 170, 172, 174, 212, 213, 217, 218, 219, 220, 221, 222, 228, 256, 261, 288, 337
United States Air Force, 172, 193, 220, 287, 288
United States Army, 301
United States Fish & Wildlife Service, 280
United States Forest Service, 260
United States Geological Survey (USGS), 262, 263, 264
United States Navy, 44, 121, 123, 306, 307, 308, 309, 310
University of Utah, 321
USS *Enterprise*, 191
USS *Lexington*, 191
USS *Monaghan*, 193
USS *Saratoga*, 191
USS *Ward*, 189
vaccines, 2, 26, 27, 28, 29, 30, 31, 32, 103, 269
Valeriano, Antonio, 127, 130
Van Allen radiation belts, 293, 294, 296
Vega (star), 127
Vibram, 56
Vikings, 176, 177, 179, 180
Virgin of Guadalupe, 126, 127, 129, 132
viruses, 30, 31, 103, 268
vitamins, 7, 27, 203, 205, 206
Walsegg, Count Franz von, 198
Warren, John, 75

waterspout, 45
Watts, Earl, 337, 338
Wayne, John, 321, 323
weather balloons, 170, 174
Wechsler, David, 301
Weems, T. Jeffrey, 340, 341, 342
Weidner, Jay, 116
Welling, Tom, 321
Westall '66 UFO, 168, 169, 170, 171, 173, 174
whale strandings, 307, 308, 309, 310, 311
whales, 306, 307, 308, 309, 310, 311
wheat, 325, 326, 327, 329
wheat allergies, 327
White, Ed, 220, 221
Wi-Fi, 13, 14, 18, 214
Wikipedia, 60
Winfrey, Oprah, 5
Winston-Salem, NC, 260
Wired (magazine), 116
Wiseman, Richard, 24
witchcraft, 245, 247, 248
Wizard of Oz, The (movie), 315, 318
Wood, Evelyn, 274, 277

Woods Hole Oceanographic Institution, 308, 310, 311
World Health Organization (WHO), 13, 14, 16, 283, 284
World War I, 301
World War II, 73, 120, 121, 122, 123, 124, 147, 148, 169, 192, 240, 251, 279, 287, 288
Wündt-Hering illusion, 334
Xinhua News, 35, 36
Xinmin Weekly, 37, 38
x-rays, 15, 50
Yali, 124
Yonaguni Monument, 231, 232, 233, 234, 235, 236
young earth creationism, 161, 162, 163, 164, 165
YouTube, 41, 68, 71, 157, 264, 331
Yuma, Arizona, 176, 177, 179
Zaortiga, Bonanat, 129
Zeitgeist (film), 107, 108, 109, 110, 111
Zheng He, 178
Zionism, 117
Zöllner illusion, 334
Zumárraga, Juan de, 126, 127, 128, 129, 131